Sex, Gender, and Science

Sex, Gender, and Science

Myra J. Hird
Department of Sociology
Queen's University, Kingston, Ontario, Canada

First published 2004 by
PALGRAVE MACMILLAN
Houndmills, Basingstoke, Hampshire RG21 6XS and
175 Fifth Avenue, New York, N.Y. 10010
Companies and representatives throughout the world.

PALGRAVE MACMILLAN is the global academic imprint of the Palgrave
Macmillan division of St. Martin's Press, LLC and of Palgrave Macmillan Ltd.
Macmillan® is a registered trademark in the United States, United Kingdom
and other countries. Palgrave is registered trademark in the European
Union and other countries.

ISBN 1–4039–2176–8 hardback
ISBN 1–4039–2177–6 paperback

This book is printed on paper suitable for recycling and made from fully
managed and sustained forest sources.

A catalogue record for this book is available from the British Library.

Library of Congress Cataloging-in-Publication Data

Hird, Myra J.
    Sex, gender, and science/ Myra J. Hird.
        p.cm.
    Includes bibliographical references and index.
        ISBN 1–4039–2176–8 (alk. paper) – ISBN 1–4039–2177–6
    (pbk.: alk. paper)
        1. Feminist theory. 2. Sex. 3. Sex differences. 4. Sex differences
    (Psychology) 5. Body – Social aspects. I. Title.

HQ1190.H567 2005
305.3—dc22   2004048941

10 9 8 7 6 5 4 3 2 1
14 13 12 11 10 09 08 07 06 05

Printed and bound in Great Britain by
Antony Rowe Ltd, Chippenham and Eastbourne.

Contents

List of Illustrations

Acknowledgments

I thank my commissioning editor, Briar Towers, for her encouragement and enthusiasm to complete this project. I also thank Jen Nelson for all of her helpful and patient editorial assistance, and Veena Krishnan for copy editing the manuscript. A number of people have helped, through conversation and support, the completion of this book, including: Ciaran Acton, John Brewer, Nigel Clark, Patricia Donald, Jenz Germon, Judy Haschenburger, Stevi Jackson, Stephen Katz, Mary Kerr, Jane Kubke, Liam Lynch, Fiona Lytle, Barbara Marshall, Cillian McBride, Lois McCammond, Deborah Parnis, George Pavlich, Sasha Roseneil, Tam Sanger, Sue Scott, Rhonda Simms, Lisa Smyth, Liz Stanley, Nicola Yeates, and the students in my 2002 module 110SOC329 at Queen's University, Belfast.

I dedicate this book to Anth who is my best friend 'even though' and to Inis, who has taught me how ridiculous human sexual reproduction really is.

Parts of Chapters 1 and 4 are reprinted by permission of *Sociological Research Online* from Hird, M.J. 'From the Culture of Matter to the Matter of Culture: Feminist Explorations of Nature and Science' *Sociological Research Online*, 2003, 8(1). http://www.socresonline.org.uk/8/1/hird.html. Parts of Chapter 4 are reprinted by permission of the University of Toronto Press from Hird, M.J. 'New Feminist Sociological Directions' *Canadian Journal of Sociology*, 2003, 28(4): 447–462. Parts of Chapter 5 and 8 are reprinted by permission of Taylor and Francis Group plc (http://www.tandf.co.uk) from Hird, M.J. 'Re(pro)ducing Sexual Difference' *Parallax*, 2002, 8(4): 94–107. Parts of Chapters 4 and 8 are reprinted by permission of Sage Publications Ltd. From Hird, M.J. 'Naturally Queer' *Feminist Theory*, 2004, 5(1): 133–137 and Hird, M.J. 'Chimerism, Mosaicism and the Cultural Construction of Kinship', *Sexualities*, 2004, 7(2): 225–240. Parts of Chapters 2 and 7 are reprinted by permission of Sage Publications Ltd. from Hird, M.J. 'Gender's Nature: Intersexuals, Transsexuals and the 'Sex'/'Gender' Binary' " *Feminist Theory*, 2000, 1(3): 347–364. Parts of Chapter 7 are reprinted by permission of Palgrave Macmillan Press Ltd. from Hird, M.J. and Germon, J. 'The Intersexual Body and the Medical Regulation of Gender' In Backett-Milburn, K. & McKie, L. (eds) *Constructing Gendered Bodies*. London: Palgrave, 2001, pp. 162–78.

Figure 3.1 is reprinted from Moore, K. and Persaud, T. (1998) *Before We are Born: Essentials of Embryology and Birth Defects*. 5th Edition. Philadelphia, PA: Saunders Company, p. 18, with permission from Elsevier. Figure 3.2 is reprinted from Oudshoorn, N. (1994) *Beyond the Natural Body: An Archeology of Sex Hormones*. London: Routledge, p. 30 by permission of Taylor and Francis (original from J. Freud (1936) "Over Geslachtshormonen," *Chemisch Weekblad 33* 1(3):1–14). Figure 3.3 is reprinted by permission of the publisher from *The Century of the Gene* by Evelyn Fox Keller, p. 24, Cambridge, Ma.: Harvard University Press ©2000 by the President and Fellows of Harvard College. Figure 7.1 is reprinted from Fausto-Sterlng, A. (2000) *Sexing the Body: Gender Politics and the Construction of Sexuality*. New York: Basic Books, p. 59 by permission of The Perseus Books Group.

The author and publisher have made every attempt to contact copyright-holders. If any have inadvertently been overlooked, the publisher will be pleased to make the necessary arrangement at the first opportunity.

1
Introduction

How should explanations be possible when we turn everything
into an image, our image!

(Nietzsche 1974: 172)

Introduction

I have been researching and teaching sexual difference theory for several years. As a sociologist, I have been trained to analyze sexual difference as the product of social construction; and as a feminist, I have
learned to look for the ways in which gender and power work conterminously to produce political, economic, social, and scientific understandings of physical (material) differences between females and males
as natural, immutable and informing.

My experience with teaching is that a concern with materialism is not
limited to feminist theorists, but is particularly distilled in students' discussions of sex "differences" (alternatively termed "sexual difference").[1]
Students, echoing a modern discourse infused with a lingering faith in
science, seem to accept that "gender" is to a large extent socialized, but
maintain that the object of socialization remains concrete, material bodies which can be neatly differentiated on the axis of "sex." In other
words, while my students are keen to explore the workings of "gender,"
it is only with the comfort of a fully (biologically) grounded "sex."[2] The
specter of nature, through scientific discourse, effects a continual return
to hormones, chromosomes, genitals, gonads, and sexual reproduction
as the material determinants of sexual difference, and my students seem
to offer up these libations with utter confidence.

My concern with my students' use of (biological) materiality to
explain sexual difference is twofold. First, although students rely heavily

1

on biology, most come from arts or social science backgrounds with little knowledge of science studies. Most of my students would not know a testosterone compound if I presented it to them under a microscope, let alone be able to say how it functions in the body, is created, circulated and transformed by other enzymes, or acknowledge its historical construction within scientific discourse. Yet this lack of knowledge does nothing to deter students' apparent blind faith in the power of hormones to determine such behaviors as aggression and sexual desire. Given the meticulous detail demanded of social constructionist theories to explain social phenomena, the cursory demands on science to explain the mechanisms of social behavior is unbalanced.

My second concern is that the invocation of biology limits the ways in which we are able to talk about social constructionism. Reliance upon a nebulous understanding of biology reifies a binary relationship between "sex" and "gender" such that explorations of gender are authorized upon the condition that "sex" is left largely intact. The readiness of social constructionism to deconstruct materiality through cultural criticism, dogged by the refusal of this materiality to disappear into text as well as its resilience within popular culture to explain sexual difference, has created somewhat of an impasse between biology and social constructionism. As Kerin argues "we cannot merely displace the force of scientific schemes by analyzing their cultural conditions of emergence" (1999: 101). Kirby's more acerbic observation crystallizes the problem: "the radical purchase of deconstruction's self-consciously awkward and repetitive insistence that the text is everywhere present, has been dissolved in the mantra of its repetition" (1999: 20).

This book has two major aims. The first aim is to outline the social study of science and nature, of which feminism is a major player, specifically in relation to "sex," sex "differences," and sexuality. Along with a host of feminist scholars of the social construction of scientific knowledge, I argue that Western understandings of "sex" are based less upon an actual knowledge of sex "differences" rooted in morphology than in a cultural discourse that emphasizes sex dichotomy rather than sex diversity. The second aim is to draw upon a loosely configured group of analyses alternatively termed "neo-materialism" (Braidotti, 2000) or "new-materialism" (Sheridan, 2002; Wilson, 1998) to further contest cultural assumptions about "sex" and sex "differences." New materialism marks a momentous shift in the natural sciences within the past few decades to suggest an openness and play within the living *and* nonliving world, contesting previous paradigms which posited a changeable culture against a stable and inert nature. I suggest these transformations within the natural sciences might be of interest to feminist social scientists who increasingly find themselves (often through "the body") grappling with issues involving life and matter. On the whole, while feminism has cast light on social and cultural meanings of sexual difference, there seems to be a hesitation to delve into the actual physical processes through which stasis, differentiation and change take place. That is, there is a paucity of feminist studies that analyze how physical processes might contribute to feminist concerns such as "the body" and sex "differences." Recent studies suggest an enthusiasm on the part of feminist theory to revisit the issue of sex "differences," directly through biological science and throughout the book I sketch how some feminist scholars are working with matter to create science–literate analyses.

The present chapter presents a review of the diverse ways in which feminist theory has explored the cultural constructions of materiality. This review is necessarily partial: rather than provide a complete exegesis on the subject, my interest is in framing the rationale behind the general shift toward feminist exploration of materiality itself.

The culture of matter

Feminist theory has been keenly involved in the project of uncovering what I have come to term the *culture of matter*, using social constructionism as its primary theoretical tool. That is, feminist scholarship focuses on developmental and cultural aspects of identity formation and negotiation, figuring sex "differences" as that which is compelled through discourse to "be" sexual difference. The contemporary feminist focus on the culture of matter was initially prompted by the recognized need to critique theories of materiality that emerged within political, economic, and social discourses during the eighteenth century (e.g. sociobiology), which began to use science as a key source of evidence for "solutions to increasing questions about sexual and racial equality" (Schiebinger, 1993: 9). These discourses cohered around the institutionalization of sexual differences between female and male nonhuman and human animals. In response, feminist critique turned its gaze toward five main areas, producing comprehensive and detailed analyses of scientific reports.

The first area, often termed feminist science studies, largely concerns the place of women in science. A number of texts (Haraway, 1989; Hubbard, 1989; Keller, 1983; Kohlstedt and Longino, 1997; Mayberry, Subramaniam, and Weasel, 2001; Small, 1984) have examined this topic, arguing that female scientists have had to particularly struggle against "masculinist" disciplines particularly in the natural sciences. Some analyses (Ainley, 1990) survey statistical trends of female students and academics in various science disciplines. Other analyses are more interested in exploring the working lives of women biologists such as Keller (1983) and Lancaster (1989, 1991); primatologists such as Altmann (1980); Fedigan (1984); Haraway (1989); Hrdy (1974, 1981, 1986, 1997); Rowell (1974, 1979, 1984); Small (1984), and Zihlman (1985); entomologists and astronomers (see Schiebinger, 1989); mathematicians (Henrion, 1997); physicists (Barad, 2001; Keller, 1977; Wertheim, 1995); and engineers (Meilwee and Robinson, 1992). These diverse studies all emphasize how women affect, and are affected by, the culture of science. And indeed, the feminist movement must be accorded credit in establishing women in science at all.

Some of these works have gone on to argue that women have historically been marginalized from all processes of scientific endeavor, which, coupled with the argument that women view and engage with the world differently than men, has prompted female scientists to approach scientific questions from a less mainstream and more creative perspective, challenging fundamental assumptions about issues such as objectivity and the "natural" inferiority of women. For instance, Evelyn Fox Keller (1983) suggests that Barbara McClintock's *Feeling for the Organism* was produced by her greater openness to alternative accounts of genetics, because she "escaped some of the psychosocial indoctrination received by her male peers" (Hawkins, 1998: 163). The argument that women approach nature and science questions from a fundamentally different perspective is particularly distilled in theories of science as social knowledge, and raises the question of the possibility of a distinct female epistemology of science (see Hankinson-Nelson and Nelson, 1996; Hubbard, 1989; Longino, 1990; Schiebinger, 1989; Stengers, 1997, 2000). Again, these analyses emphasize the social, political, and economic features of women in science, rather than the actual material objects that female scientists study.[3]

Adopting many of the premises of "feminine science", the third area of materiality that feminists have theorized is eco-feminism (for a useful summary see Soper, 1995). Eco-feminism comprises a diverse range of approaches, both theoretical and practical, to the impact of human animals on living and nonliving matter. Some of these theories utilize arguments for a distinct feminist epistemology, as the means by which women are better able to live within the world without destroying it. One of the more positive aspects of eco-feminism is the "recognition of nature in the 'realist' sense ... nature as matter, as physicality: that 'nature' whose properties and causal processes are the object of the biological and natural sciences" (Soper, 1995: 132). This conception of nature focuses on processes of living and nonliving matter that are independent of human activity, a theme I will return to later in the book.

A fourth area of feminist attention is the relation among nature, science, and technology, with reproductive technologies and cybercultures topping the list of recent focus. Many feminist considerations of the future of sexual difference focus on the impact biotechnologies might have on conceptualizations of embodiment and materiality. A host of feminist contributions concentrate either on the supposed rent of maternal embodiment from women through cloning and other reproductive technologies (Murphy, 1989; Overall, 1989; Sawicki, 1999; Spallone and Steinberg, 1987; Weir, 1998 – see Donchin, 1989, for a useful review) and increasingly genetics (Franklin, 1995, 2000), or the transformative possibilities which technologies, through cyborgean technobodies, potentially offer (Braidotti and Lykke, 1996; Broadhurst Dixon and Cassidy, 1998; Featherstone and Burrows, 1995; Gray, 1995; Haraway, 1991; Jonson, 1999; Plant, 1997).

The fifth area of feminist analysis concerns the social construction of scientific knowledge. Briefly, these studies begin their analyses from the perspective that scientific 'facts' are socially mediated and can only be understood within their particular social and cultural milieu. Chapter 2 of this book examines the development of science in the eighteenth and nineteenth centuries in order to argue that science enabled Western social, economic, and cultural discourses to emphasize sex "differences" rather than sex diversity, or intrasex (within sex) differences and intersex (between sex) similarities. Chapter 3 goes on to examine feminist literature on the social construction of scientific knowledge concerned with what is often termed the "essence" of sex and sex "differences": gonads, hormones, chromosomes, and genes. The chapter analyzes scientific accounts of sexual difference based upon skeletons, hormones, chromosomes, egg and sperm activity, and animal behavior (the list of supposed sex "differences" also includes sexual reproduction, the critical analysis of which is so extensive as to merit its own chapter – see Chapter 5). The chapter then explores the ways in which science and culture often work conterminously to reinscribe sexual difference on to the human body.

For instance, through an exploration of the history of chromosome study, Chapter 3 explores the ways in which *a priori* scientific and cultural inscriptions of sexual "difference" served to direct analyses toward the confirmation of sexual dichotomy, despite abundant evidence of sex diversity. The chapter's analysis of the inscription of cultural notions of sex "differences" on to the physical processes of egg and sperm activity provides the basis for an analysis of human gonads as another site deployed in the social construction of sexual "difference". As Fausto-Sterling suggests, behind debates about sexual reproduction "lurk some heavy-duty social questions about sex, gender, power, and the social structure of European culture ... In the work of the established evolutionary biologists, past and present, talking about eggs and sperm gives us permission to prescribe appropriate gender behaviors" (1997: 54, 57). The chapter provides a similar analysis of the development of research on hormones, which were "created" through the lenses of both sex "differences" and heteronormativity. However successful feminist arguments concerning the social construction of gender have been within academia and the public in general, there remains a persistent and robust recourse to a biological notion of sexual "difference" based upon often cursory notions of testosterone levels or X and Y chromosomes. For this reason, Chapters 2 and 3 employ feminist theory to critically review each of these "facts" of "sex" with a view to highlighting the mechanisms through which scientific knowledge is constructed.

In sum, each of these five areas of feminist research illuminate, in different ways, how social and cultural discourses inform understandings of the science and "nature" of "sex" and sex "differences." The fifth area of study particularly focuses on the social *production* of scientific facts about sex "differences." Does this suggest that feminists should eschew scientific studies of sex and sex "differences" altogether? Our contemporary understanding of the concepts "sex" and "sexual difference" are founded upon the acceptance of scientific "facts." But if these facts are culturally mediated, can, or should, we use science at all to explore "sex" and "sexual difference"? Would we not be better off remaining entirely within the cultural domain, explaining these concepts as entirely socially constructed?

The matter of culture

The fifth area of feminist critique described above convincingly argues that the "naturalness" of bodily materiality is socially mediated, and raises significant questions about the relationship between the cultural and the physical. However, these analyses largely tend to open up science to the social, leaving the actual materiality of organic and nonorganic matter intact. For this reason, the book centrally argues that whereas "the body" is meant to signify nature, what is actually being analyzed are sites at which culture meets nature. Thus, "the body" does not actually signify materiality in its own right, but in fact resignifies culture.

Discussions of materiality within sociology and much of feminist theory tend to be anchored by two critical assumptions (Soper, 1995). First, the constitution of matter is largely figured as inert, stable, concrete, unchangeable, and resistant to sociohistorical change. Second, science is viewed as one of the strongest bastions of patriarchy. Each of these assumptions continue to shape the ways in which materiality is studied, which in turn impacts on the parameters set for exploring "sex" and sex "differences."

Conceptions of matter

For the most part, feminist scholars seek to analyze women's experiences within social, cultural, and political milieus using a social constructionist framework which ultimately argues for the impossibility of analyzing "sex," and "sexual difference" from outside of the social practices that create and sustain them. In other words, according to social constructionist analyses, "sex," and "sexual difference" are historical artifacts rather than biological phenomena, and thus cannot be "neutral" objects of scientific investigation. This view is perhaps most powerfully forwarded by Michel Foucault (1979, 1980), who argues that power, knowledge, and truth are coextensive. That is, what we understand to be the "truth" of concepts such as "sex" are structured by power and knowledge paradigms to an extent that precludes the basis of scientific discourse about the neutrality and objectivity of science to uncover the "facts" of the material world. Timothy Murphy (1997) acknowledges that while knowledge, power, and truth can be nothing other than coextensive, he maintains that this does not disqualify scientific inquiry altogether: "the relevant question is not, therefore, whether science can achieve some neutral methodology and goals – that have, for example, no disciplinary functions – but whether particular forms of discursive and scientific power are objectionable in their conception and effects" (1997: 62). For this reason, it is vital to subject the natural sciences to scrutiny.[4]

For example, much has been written within feminism on eating disorders and the body, including the social construction of dieting, fitness, beauty, and the patriarchal system that regulate women's relationships with their own bodies (Bordo, 1993; Orbach, 1986). Despite the enormous number of feminist analyses on the gendered construction of eating disorders, "these analyses consider the cellular processes of digestion, the biochemistry of muscle action, and the secretion of digestive glands to be the domain of factual and empirical verification ... only a certain understanding of the body has currency for these feminist analyses, an understanding that seems to exclude 'the biological body' " (Wilson, 1998: 52).

In another example, David Halperin discusses the potential consequences of scientific investigations into genetic contributors to homosexuality (this topic is discussed further in Chapter 3). Halperin concedes that "if it turns out there actually is a gene, say, for homosexuality, my notions about the cultural determination of sexual object-choice will – obviously enough – prove to have been wrong" (1990: 49). Halperin's concession is based upon a dichotomous relationship between "nature" and culture; that if something is "natural" it cannot be cultural and vice versa. But as Murphy (1997) points out, even if scientific analyses did find a genetic component for homosexuality (and I will contest this possibility later in the book), homosexual erotic interest will still *necessarily* have a cultural component because the expression of homosexuality will be entirely different within the cultures that accommodate homosexuality than in cultures that do not.

The strict division between "nature" and "culture" evinced in many "essentialist" versus "constructionist" debates thus does not make sense insofar as it artificially separates two aspects of what ultimately produces behavior. Oyama (2000) argues that discussing biological determinism and social constructionism in opposing terms replays the familiar nature–nurture debate such that the "cause" of some phenomenon is argued to be the result of *either* nature *or* nurture. Gray similarly argues that "the dichotomous view of development, that environmental factors produce acquired behavior and genetic factors produce innate behavior, leads to the erroneous view that if a behavior is present at birth it did not require any environmental factors to develop, and that it will not be changed by subsequent experience" (1997: 389). Part of the problem is that data that purports to show some sort of genetically sex dimorphic behavior also tends to be the same data that could be equally argued to be entirely cultural in explanation (Allen, 1997).

Vicki Kirby argues that contemporary critical analyses' insistence that the target of scrutiny is the discursive effects of objects, and not the object themselves, belies a construction of materiality as "rigid, prescriptive" and opposed to "cultural determinations that are assumed to be plastic, contestable, and able to invite intervention and reconstruction"

(2001: 54). Consequently, as we have seen, when feminists study materiality, it tends to be in terms of how humans (such as scientists) interact with materiality, as though there is no outside of, or beyond, the cultural context. Anne Witz explains:

> Feminist sociologists have, for the most part, written against the grain of corporeality, in the sense of a fleshy materiality, in order to fill out the absent, more-than-fleshy sociality of women traditionally repressed within sociological discourse. And for good reasons. Precisely because they were sociologists, they did laterally for women what masculinist sociology had formerly done for men, and men alone: they retrieved them from the realm of the 'biological', 'corporeal' and 'natural' and instated them within the realm of the 'social'. (2000: 4)

The difficulty with social scientific and cultural analyses of the representation of matter is that "providing a social explanation ... means that someone is able in the end to *replace* some object pertaining to nature by *another one* pertaining to society, which can be demonstrated to be its true substance" (Latour, 2000b: 109). This produces a recursive return to sociality and away from the material object of study. But as Judith Butler acknowledges:

> it must be possible to concede and affirm an array of 'materialities' that pertain to the body, that which is signified by the domains of biology, anatomy, physiology, hormonal and chemical composition, illness, weight, metabolism, life and death. None of this can be denied. (1993: 66)[5]

This particular problem, that social constuctionist accounts of "sex" and sex "differences" tend to be limited to cultural analyses, presents a particular challenge. Most social theories invoke "bodily materiality" with little knowledge of evolution, biology, anatomy, chemistry, or physics. This lack of knowledge about – to borrow from John Brockman and Katinka Matson – *How Things Are* (1996), sets necessary limits on discussions of materiality, notwithstanding the most extreme deconstructive efforts.[6] As a first step, then, we must challenge ourselves to become sufficiently literate in the natural sciences to contemplate the contribution of this knowledge to feminist theory. This book considers the "nature" of matter specifically in relation to "sex," sex "differences," and "sexuality." Rather than provide a hard-and-fast definition of "culture" and of "nature," I will revisit these concepts throughout the book, with a view to exploring why they are difficult to define especially in relation to "sex" and "sexual difference."

If, as Elizabeth Wilson argues, feminist critiques of the *culture of matter* "have fallen into old and familiar patterns [then one way this] ubiquitous gravitation to culture" (1996: 50) might be averted is by studying the *matter of culture*. Current and future debates within feminist theory might productively engage with materiality, and I see in the general shift toward new materialism a desire to formulate new questions for feminism. For instance, new materialism presents a general paradigm shift in our understanding of the relationship between nature and culture. Taking into account the idea that matter possesses its own "immanent and intensive resources for the generation of form from within" (De Landa, 2000) might help us to think about materiality without the usual accompaniment of essentialism, where matter is understood as an inert container for outside forms. These observations of nature might aid feminist reflections on theories of sex complimentarity such as the division of labor in society and the "public/private" divide. Nonlinear biology affords an opportunity to explore the issue of boundaries from a different perspective. Boundaries involve a number of issues, from the integrity of bodies as single organisms, the taxonomy of sexuality (homosexuality/heterosexuality), to the differentiation of humans from other primates, and indeed living and nonliving matter in general. As Sarah Franklin notes "trading organismic distinction for pan-species genetic information flow pulls the rug out from under the sex/gender system as we know it" (1995: 69).

"Sex" is another area within which a great deal of boundary work is done. Sharon Kinsman argues:

> Because most of us are not familiar with the species, and with the diverse patterns of DNA mixing and reproduction they embody, our struggles to understand humans (and especially human dilemmas about 'sex', 'gender' and 'sexual orientation') are impoverished ... Shouldn't a fish whose gonads can be first male, then female, help us to determine what constitutes 'male' and 'female'? Should an aphid fundatrix ('stem mother') inform our ideas about 'mother'? There on the rose bush, she neatly copies herself, depositing minuscule, sap-siphoning, genetically identical daughters. Aphids might lead us to ask not 'why do they clone?' but 'why don't we?' Shouldn't the long-term female homosexual pair bonding in certain species of gulls

help define our views of successful parenting, and help [us] reflect on the intersection of social norms and biology? (2001: 197)

Against all analyses that use "nature" to argue against sexual diversity, Bruce Bagemihl convincingly argues that "natural systems are driven as much by abundance and excess as they are by limitation and practicality" (1999: 215). Bagemihl presents a comprehensive catalogue of homosexual, transgender, and nonreproductive heterosexual behavior in animals that defy the traditional homosexual/heterosexual boundary. Grosz notes that "the particular characteristics defining an insect species ... are always in excess of their survival value. There is a certain structural, anatomical or behavioral superabundance" (1999b: 280). Gay parenting, lesbianism, homosexuality, sex-changing and other behaviors in animals are found, *in abundance*, in strong species or ecosystem.

How might we interpret the lack of diversity in human culture as compared with the enormous diversity in the nonhuman living organism world? We might choose to interpret our lack of diversity in terms of the uniqueness of our species, or we might understand this lack of diversity in terms of culture and politics. Patricia Gowaty argues:

> The evolutionist in me argues with essentialist feminists and likewise with some evolutionary biologists that attention to fixed, invariant, and universal differences among women and men is likely to miss the mark most of the time ... First, the diversity and variation among individuals is one of the most impressive of human 'universals', and anyone seeking a unified theory of human nature must account for the impressive variation among, between, and within individuals ... It seems to me that cultural variation in humans may be an excellent example of how some universal selection pressures acting on the interactions of women and men could have led to the enormous within- and between-cultural variation that characterizes humans (1997b: 6–7).

Finally, new materialism and nonlinear biology provide an occasion to explore debates within feminist theory about whether or not technology is helpful or a hindrance to women. In these discussions, it is worth bearing in mind that life itself is, and has always been, "technological" in the very real sense that living organisms incorporate external structural materials into their bodies (Margulis and Sagan, 1997). Moreover, much of the "new" technologies (including reproductive technologies, cloning, and stem cell research) was already perfected millions of years ago by other living matter on this planet.[7]

In sum, a social constructionist approach to science need not eschew consideration of materiality. Developments in new materialism and nonlinear biology suggest new ways of analyzing "sex" and sex "differences." By largely confining analyses to "gender," feminist theory may have missed some more radical questions. For instance, materialy speaking, it is not at all obvious why human beings have sex. Most species on this planet are not sex dimorphic and scientists are more challenged to explain sex "differences" (and, as a consequence, sexual reproduction) than they are to explain the much more prolific abundance of sex diversity on this planet, including intersex, transsex, transspecies sex and reproduction through mitosis, fusion and so on. Moreover, by assuming that "sex" is stable and immutable, we may miss useful means of analysis such as evidence of transsex in nonhuman species to inform analyses of transsex in human cultures.

Science and patriarchy

While the first assumption is based upon the social constructionist conception that analyses of matter are necessarily limited to discourse, the second, related, assumption is that the primary means through which the study of matter has been accessed (science) is principally a tool of patriarchy. Consequently, feminists often approach analyses of matter both reluctantly and negatively. As Elizabeth Grosz notes, "nature has been regarded primarily as a kind of obstacle against which we need to struggle" (1999b: 31). Thus, feminist analyses focus on reproductive technologies, premenstrual syndrome, menopause and birthing technologies in often-negative terms (Balsamo, 1996). In part, the negative inflection of these analyses is guided by a commitment to the feminist political project of equality, which is keenly sensitive to any natural science inclinations toward sex, sex "differences," and "the nature of things." Social scientists tend to focus on the social because it is viewed as historically dependent, and thus changeable.[8]

This second assumption presents a second challenge to analyses of "sex" and "sexual difference": to move away from a notion of matter as an inert, largely negative ontology whose only representative medium is masculinist science, and toward a more positive notion of matter as open-ended and playful. The revived interest in feminist science studies indicates a movement beyond the political position that science serves only to endorse the claims about women's "ontology" that the political project of feminism set out to challenge. Here we find a reconsideration of the notion of bodies as the excluded "other" to masculinist representation (see Colebrook, 2000a, b). Using both Deleuzian (Deleuze and Guattario, 1987; Deleuze, 1994) and Derridean (1978) theories in tandem with science literate analyses of matter, feminist concerns such as "the body" and sexual difference are explored.

Signposting the narrative

Throughout the book I focus on "sex" by which I mean the cultural dichotomy established between "females" and "males." I must attach quotation marks to each of these concepts in order to emphasize their cultural genesis, rather than their definition as rooted in materiality. Indeed, the book strongly argues that "sex" – defined in terms of a dichotomy – only makes sense within the cultural–political framework of its associated concept "sexual difference." In this book, "sexual difference" (or sex "differences") is defined in terms of "sex complementarity" – the emphasis on differences between females and males rather than similarities. In reviewing the literature, I am concerned every time I read the word "opposite" in reference to either "sex" or "sexual difference." Parts of this book will also explore aspects of "sexuality" inasmuch as this concept is anchored by the concepts of "sex" and "sexual difference," although the major focus of the book will remain on "sex" and "sexual difference." I have chosen to focus on the term "sex" rather than "gender" because I want to emphasize those analyses that invoke a material notion of the relationships between "females" and "males," rather than cultural, social, or political associations.

In this book, Chapter 4 takes up the challenge of introducing a positive notion of matter that emphasizes open-endedness, activity, and playfulness. The chapter begins by providing a concise review of the major rationale of evolutionary theory, followed by an introduction to the notion that the physical sciences are moving away from traditional evolutionary theory's understanding of nature as a stable, monolithic, and inert entity, toward a conception of nature as a complex open system subject to emergent properties. In such analyses, nature is far from inert; emergent hybridizations are not solely the product of human agency, but are indigenous to networking open systems. New materialism emphasizes that if "nature" is to retain any meaning at all it must signify superabundant diversity. Thus, I argue that science aims

not to distill the vast variation found in nature to a simple, single explanation of "reality," but rather to normalize diversity (Ferguson, 1997).
The chapter concludes by signaling the momentous shift in the natural
sciences within the past few decades to suggest that there is openness
and play within the living *and* nonliving world, contesting previous paradigms which posited a changeable culture against a stable and inert
nature. Indeed, one of the central foci of this book is the propensity of
nature to produce sex diversity rather than the dichotomous notion of
sex prevalent in cultural discourses. The chapter will also consider how
exploring matter might evacuate feminist theory from a largely negative
nostalgia (see note 3) and provide a positive theory of corporeality.

The remaining four chapters of the book provide explorations of "sex"
and "sexual difference" using new materialism and nonlinear biology.
Chapter 5 focuses on the nonlinear evolution of human sex. The ability of women to sexually reproduce is the most frequent and powerful
signifier of sexual "difference" in Western culture. All feminist texts
contest the oppression experienced by women through cultural assumptions about sexual difference, but few texts actually contest the basis of
the claim that sexual difference exists. This chapter challenges the
a priori acceptance within feminist theory of sexual difference based
upon sexual reproduction. This acceptance is based on three widely held
assumptions: (1) sexual reproduction is the most common form of
reproduction among living matter; (2) sexual reproduction has an
evolutionary purpose; and (3) the human body itself is sexually
differentiated. Drawing upon data from new materialism and nonhuman animal studies, this chapter challenges each of these assumptions.
The chapter argues that the current recourse to "the body" based upon
reproductive function selectively attends to one aspect of "materiality" –
that is, human bodies (like all other living organisms) engage in
constant and varied reproduction, and only a small proportion is sexual.
The chapter concludes by arguing that "nature" has been erroneously
called upon to support the "truth" of sexual difference based on sexual
reproduction.

The aim of Chapter 6 is to introduce what is known as the "quiet
revolution" (Bagemihl, 1999) in biology – that is, the diverse range of sex
"differences," and sexual activities in strong species and ecosystems. The
chapter reviews how heteronormative assumptions about "sex," gender,
and sexuality have influenced traditional biology to erase and silence
sex diversity among living matter. This chapter argues that the vast
majority of species on this planet display a diverse range of sexes and
sexual activity, and will document what some might argue are lesbian

and gay parenting, lesbianism, homosexuality, sex changing, and other behaviors in animals, plants, fungi, and bacteria.

Chapter 7 applies the same principles developed in the previous chapter to the study of sex, sexual difference, and sexuality in human animals. By this stage, we know that the human body is intersexed, since all cells, except those that make up sperm and eggs, have "female" and "male" chromosomes. We also know that over half of our genetic inheritance comes from our mothers (through mitochondria). Nevertheless, cultural notions of sex and sexual difference still maintain that when we take the human body as an autonomous entity, it is clearly sexually differentiated.

People with intersex conditions provide a valuable opportunity to explore the relationship between "sex" and "gender," as well as the designation of meaningful categories of sexual "difference". Far from being a rare statistical anomaly, about two in every hundred people are intersex. Given the superabundance of sex diversity among animals, plants, fungi, and bacteria, the prevalence of intersex in human beings should not be particularly surprising. As such, it is particularly relevant to reflect upon the current silence surrounding Western society's attempts to eradicate intersex from the human population. Drawing on my own research on intersex, I review some of the more prevalent intersex conditions, as well as current medical protocols. The chapter concludes by analyzing the implications of intersex for feminist debates about sexual difference.

The current "management" of intersex in Western culture reveals that the *authenticity* of sex resides not on, nor in the body, but rather results from a particular nexus of power, knowledge, and truth. People with intersex conditions' experiences of "sex" challenge Western society to the extent that society is predicated on the sex/gender binary to operate. To effect the incorporation of an intersexual person surgically assigned as "female" involves a determination as to the constitution of femaleness. Any definition of "woman" that retains any corporeality must be able to define that corporeality and this is exactly where the problem begins in definitions based on "sex."

Finally, Chapter 8 brings together the evidence harnessed in the previous chapter to consider why feminists need not "reject" science studies *tout court*. It will argue that bodies are important and certainly "material," but not necessarily in ways which justify continued emphasis on sexual difference. To illustrate both the evolutionary history of sex diversity, and its prevalence within the living world, this chapter examines the sex life of bacteria. As the oldest surviving living matter on this planet, and as the evolutionary origin of all living organisms, bacteria actualize "sex" in its ultimate diversity, defying cultural understandings of sexual dimorphism and sexual reproduction. I use the example of bacteria to emphasize the point that human cultural regulatory discourses surrounding sexual difference are particularly limited compared with the sex diversity evident in nature. The chapter concludes with a consideration of how advances in new materialism aid feminist reflections upon contemporary issues including the Human Genome Project, the Visible Human Project, reproductive technologies, cloning, and xenotranplantation. Finally, the chapter examines the implications new science studies might have on feminist theory and praxis.

Suggested readings

Margulis, L. and Sagan, D. (1995) *What is Life?* Berkeley, CA: University of California Press.
Schiebinger, L. (1999) *Has Feminism Changed Science?* Cambridge, MA: Harvard University Press.

2
Making Sex, Making Sexual Difference

Similarity is the shadow of difference.

(Ridley, 2003: 7)

Introduction

This book would have no basis for discussion if contemporary Western society did not take "sex" and "sexual difference" for granted. Scholars of history have for some time argued that "sex" and "sexual difference" have undergone significant shifts in meaning. This chapter provides a historical account of the *making* of "sex" and "sexual difference." Using feminist research, the chapter focuses on what I have termed the "culture of matter" – that is, how culture has produced a discourse of "sexual difference" and complementarity, rather than some other discourse such as sex similarity. The "story" of sex "differences" is largely a story of the emphasis of difference rather than similarity, of intrasimilarity and interdifference. But, as Matt Ridley's quote above suggests, the shift to emphasizing supposed differences between women and men carries the specter of its social construction: it is equally plausible to discuss sex similarity as it is to discuss sex difference.

Pre-Enlightenment and the discourse of "one sex"

It is testament to the hegemonic power of discourse that it is often difficult to imagine our ancestors living with a radically different understanding of sex. But in pre-Enlightenment Western society, such was the case. In contemporary society, we typically refer to "sex" as the morphological and biological differences between females and males, and "gender" as any cultural differences. However, in premodern society,

17

"sex" did not hold such foundational status. As Londa Schiebinger argues "sex before the seventeenth century ... was still a sociological and not an ontological category" (1989: 8). Indeed, what we understand as "sex" today more closely resembles what, during the pre-Enlightenment period, we would term "gender." The change from "gender" to "sex" as a foundational ontology was achieved through a slow epistemic shift – not in the body itself, but in the meanings attributed to this body. This epistemic shift was made possible by the emerging discipline of science, and biology more specifically.

Since Simone deBeauvoir's classic work on biology (1949/76), a number of studies have begun to argue that, prior to the eighteenth century, women and men were considered to share one morphological body (Daston and Park 1998; Laqueur 1990; Oudshoorn 1994; Tuana 1989).[9] For instance, for the ancients, "sex" was determined by the quantity and quality of the "seed" in reproductive fluids. Indeed, while males were generally understood to originate from strong seed, a constant struggle between female and male seed was thought to ensue within each body. Any questions that the separation of sexes might invoke (if females have such powerful seed, why does she need a male in order to reproduce?) was avoided by the "one-sex economy of fluids ... in which the more potent seed is by definition the more male, wherever it originated" (Schiebinger, 1993: 40). Femininity and masculinity were determined more by close attention to signs of movement, temperament, voice and so on which indicated on which side of the one axis of "sex" any individual gravitated – active/passive, hot/cold, formed/unformed, informing/formable. That is, individuals were thought to be positioned on a single axis of "sex":

Masculinity——————————————————————Femininity

Interestingly, this single axis also applied to what, in contemporary society, has become the emblem of sexual difference: genitals. The Greek myth of Zeus depicts the father of all gods inventing interior reproduction by relocating the penis inside half of the human population. Thus, women's genitals were seen as simply male genitals displayed internally rather than externally. This idea persisted throughout the pre-Enlightenment period. Being the superior form, male bodies contained the heat necessary to "display" the penis and scrotum externally; lacking heat, female bodies bore their penis and scrota internally. The leading medical and philosophical scholars detailed the anatomical equivalence of vagina and penis, labia and foreskin, uterus and scrotum, ovaries and testicles (indeed, separate words for these body parts were only invented as a result of the two-sex model). Countless drawings, often produced from dissections, depicted the vagina as an internal penis. Only as a result of considerable controversy and political upheaval did the contemporary "two-sex" model eventually dominate scientific discourse, and "an anatomy and physiology of incommensurability replaced a metaphysics of hierarchy in the representation of woman in relation to man" (Laqueur, 1990: 5–6). During the pre-Enlightenment, genitals did not signify the founding "essence" of sexual difference – "sexual temperament" was more important an indicator of an individual's sex.

During the sixteenth, seventeenth, and eighteenth centuries what we would term "gender" held the same definitional status as our modern understanding of "sex." "Men" were defined by the characteristics of heat, strength, and rationality while "women" were characteristically defined as colder, weak, and emotional. Again, these were characteristics of degree, with men and women sharing one axis. The one axis for "sex" afforded a fluidity of movement across the gender continuum, with a large number of possible variations. As Londa Schiebinger argues "the one-sex body, because it was construed as illustrative rather than determinant, could therefore register and absorb any number of shifts in the axes and valuations of difference. Historically, differentiations of gender preceded differentiations of sex" (1993: 62).

Medical literature during this time is replete with accounts of individuals changing sex. For instance, Ambroise Pare details several stories of people's genitals changing from internal to external display. For instance, Marie became Manuel when her penis was expelled from her body when she began menstruating; a young man in Reims who lived as a girl until at the age of 14 began "frolicking" with a chambermaid and his penis was suddenly displayed outside of his body; a young man was once a girl until she jumped across a ditch and the exertion pushed her penis outside of her body: "Marie, soon to be Marie no longer, hastened home to her/his mother, who consulted physicians and surgeons, all of whom assured the somewhat shaken woman that her daughter had become her son" (Laqueur, 1990: 126. See also Daston and Park, 1998; Schiebinger, 1993).

Most of the accounts detail the change of women into men, society believing the body would always attempt to become more perfect (male).[10] Through the movement of the penis from interior to exterior, the body could express the "sex" characteristics that most suited the individual's disposition and behavior. As men enjoyed greater social and economic privileges, magistrates were most concerned to maintain structural boundaries between women and men than with the "authentic" "sex" of the individual.[11] And yet since everyone had to share one sex, individuals were freer to express variations of character which would become highly problematic once the two "opposite" sex model was adopted. As Laqueur argues, "the modern question, about the 'real' sex of a person, made no sense in this period, not because two sexes were mixed but because there was only one to pick from and it had to be shared by everyone, from the strongest warrior to the most effeminate courtier to the most aggressive virago to the gentlest maiden" (1990: 124).

Rather than demonstrating the advance of modern understandings of the body, these analyses suggest that objects do not express meaning in and of themselves, but are made meaningful in their interpretation; that we continue to superimpose dichotomies onto shades of variability. Thus, Renaissance drawings depicting the vagina as an interior penis reveal that dominant discourse, not accurate observation, determines how the body is seen and understood. Therefore, it is not that we now know the "truth" of the body: rather that "gender" discourses *are already at work* on any discussions of "sex," before they begin.[12] In short, like "gender," "sex" is an invention.

The Enlightenment and the discourse of "two-sexes"

By the nineteenth century, the understanding and practice of "sex" based upon signs of temperament, behavior, clothes, and posture was usurped by a formulation of sex as fixed, essential, and demonstrating sexual *difference*. For instance, Londa Schiebinger (1993) charts how eighteenth, nineteenth, and twentieth century European botanists attempted to find supporting evidence for the normative preference for heterosexuality, sexual reproduction, and the theory of sex complementarity. The history of botany shows a remarkable insistence on the recreation of reassuringly familiar concepts such as sexual difference among plants (despite the fact that most flowers are intersex), marital bonds between plants (the term "gamete" originates from the Greek word, *gamein*, "to marry"), active male and passive female sexuality ("male" stamens were said to have visible orgasms as opposed to the "female" pistils which showed little sexual excitement and modesty) and monogamy (even though plants reproduce through pollination which is transported via insects and air) (1993: 105). Turning from plants to animals, Schiebinger questions the classification of *Mammalia* (meaning "of the breast") when only half of this group of animals have functioning mammae, and then only for a short period of time (1993: 41). Mammae were chosen above several other possible taxonomic markers to be symbolic of women's association with nature (and at a time when politicians were attempting to convince middle-class women to breastfeed their children rather than use working-class wet nurses) whereas *Homo sapien* was chosen to associate "man" with "reason" (for more critiques of the social construction of sex "differences" in animals see Fedigan, 2001; Grosz, 1995; and Merrick, 1988).

Thus, "sexual difference" was revealed through the emerging discipline of biology, focused on the physical body as the only signifier of "sex." In terms of the genealogy of "sexual difference" then, the replacement of "gender" by "sex" as the fundamental category was the most important artifact of the Enlightenment. This is not to say that there was any one profound scientific "discovery" that overturned the "one-sex" paradigm. Indeed, this chapter will argue that sexual difference did not suddenly "reveal" itself through scientific disciplines such as medicine. Rather, slowly emerging epistemological and political shifts over a period of several centuries attributed different meaning to the same body.

The Enlightenment project was heavily dependent upon what would eventuate as a dramatic shift in epistemology. This was an overarching shift from the revelation-based knowledge of premodern society, to the scientific-based knowledge we are familiar with today. In the old system, the universe, including all that was cultural and natural, conformed to a hierarchical structure created by divine purpose. What made the Enlightenment period revolutionary was the serious challenge to the traditional social order based upon divine right rather than the free will of the people. The Enlightenment promised to distinguish fact from fiction, reason from revelation and superstition, and science from religion.

Michel Foucault outlines the consequences of this slow epistemic shift on approaches to bodies (1994a). Whereas at one time it was considered sacrilegious to tamper with the internal body, the emerging science of anatomy began to transform the body into detachable pieces, or "organs without bodies" (Schiebinger, 1993). This anatomy took the form of at first private, and then public, dissections of human and nonhuman animal bodies, and also increasingly detailed anatomical drawings in books. In contrast to pre-Enlightenment explorations, doctors were less interested in the body as a holistic whole, and much more concerned with the microfunctions of its constituent parts, each part to be classified in an overall taxonomic structure. The study of anatomy led to other emerging fields including molecular biology, biochemistry, endocrinology, neurobiology, and histology. These subfields even more sharply focused attention on microstructures and functions: cells, hormones, neurotransmitters, and so on. Foucault (1994a) points out that an important element of this epistemic shift was the emerging focus on the body revealing its secrets through visualization. Dissection literally opened up the body to scrutiny and medical scientists were able to focus on the "truth" revealed by the inside of the body rather than on the "superficiality" of the outside of the body (Schiebinger, 1993). Part of this "truth" was "sexual difference," and "sex" began to permeate throughout the body, no longer in the form of seed, heat, or humors, but in visible objects. For instance, the premodern view of the penis as anatomically expressed either inside or outside of the body, now gave way to a separate classificatory scheme: male penis and female vagina. Ovaries and testes, once considered the same organ, were also distinguished by their own names. And in an ironic twist of history repeating itself, in 1559 Renaldus Columbus claimed to have discovered the female clitoris (Schiebinger, 1993). This was a significant declaration because it suggested the "truth" of the body could be found visually and manually (through touch), and also because it further challenged the one-sex paradigm insofar as females could not have two penises – vagina and clitoris. The search for anatomical differences between women and men did not rest with the genitals and gonads. Anatomy books began to provide detailed pictorial descriptions of differences in skeletons, brains, skulls, hair, eyes, sweat, blood vessels, and so on. In this way, "sex" and "sexual difference," began to permeate the entire body and "by the 1790s, European anatomists presented the male and female body as each having a distinct telos – physical and intellectual strength for the man, motherhood for the woman" (Schiebinger, 1989: 190–191).

But the move to the contemporary "two-sex" paradigm of sexual difference would not have been achieved through this epistemic shift alone. A forceful political dimension was also necessary. If the Enlightenment project was based upon a fundamental overthrow of a social order based upon divine hierarchical privilege, then what would the new social order look like? If the Enlightenment was forged with promises of equality and justice, how was the continued subordination of women to be reconciled? As the doctor Louis de Jaucort explained "it appears at first difficult to demonstrate that the authority of the husband comes from nature because that authority is contrary to the natural equality of all people" (Schiebinger, 1993: 215). The answer lay in the very shift to biology and science as the ultimate purveyors of knowledge and "truth." Scientists and politicians alike turned their attention to "nature" to tell a new story of sexual difference. As Schiebinger argues, "an anatomy and physiology of incommensurability replaced a metaphysics of hierarchy in the representation of woman in relation to man" (1993: 6). That is, if the one-sex paradigm understood women and men to be degrees on the same axis, scientists and politicians began to see "nature" as revealing women and men to be on opposite ends of completely separate scales:

Masculinity————————————————————————————————
————————————————————————————————————Femininity

As well as the increasing number of medical books detailing anatomical differences between women and men, politicians and social critics wrote treatises that emphasized "sexual difference." For instance, Jean Jacques-Rousseau's famous novel *Emile* sought to ground his arguments about the incommensurability of women and men through biological difference. Indeed, Rousseau maintained that "a perfect woman and a perfect man ought not to resemble each other in mind any more than in looks" (Schiebinger, 1993: 226). Rousseau's books, like those of his contemporaries, were particularly influential because they were able to bring apparently new evidence to old arguments. That is, this new politics sought to maintain old hierarchies, not through notions of the divine rights of men, but through the newly emerging biological foundation of *sex complementarity*. Specifically, sex complementarity held that women and men were, biologically, better suited to different roles, and that these roles complemented each other to form the optimum living, working system. Women were to maintain the family and household while men controlled the public and political sphere. Sex complementarity maintained the gendered division of labor between private and public spheres by taking up the new sciences of biology and anatomy that were already at work emphasizing "sexual difference." In this vital way, biology, as the purveyor of stable, ahistorical, and impartial "facts" about "sexual difference," became the foundation of political prescriptions about social order. Thus Geddes was able to pronounce to the British parliament in 1889 that "what was decided among the prehistoric Protozoa cannot be annulled by an act of Parliament" (in Laqueur, 1990: 6).[13] So it was not a human-made political order that maintained women's subordination and disenfranchisement, but "nature" itself that revealed social inequality. Thus, Emile Durkheim may have been more correct than he knew when he observed that "the two sexes do not share equally in life [and that] gender difference and inequality are the by-products of modernity" (in Marshall, 2000: 19).

The contemporary sex/gender binary

In contemporary society, the conceptual division between "sex" as the biological differences, and "gender" as the social, cultural, economic, and political differences, between women and men is largely taken for granted. Although most people might not know this division as the "*sex/gender binary*," nevertheless most discussions of gender inequality eventually distill into proclamations of biological sex differences. However, within feminist theory, and perhaps the social sciences more broadly, this binary has undergone vigorous challenge for some years. For instance, in 1997, the journal *Differences* devoted an issue to questioning the continued viability of Women's Studies as a discipline, adjoining a growing "identity" debate within feminist theory. Notwithstanding a generally supportive attitude, the contributors acknowledge a general movement away from Women's Studies toward gender, gay, lesbian, and sexuality studies:

> ... women's studies has sometimes greeted uncomfortably (and even with hostility) the rise of feminist literary studies and theory outside its purview, Critical Race Theory, postcolonial theory, queer theory, and cultural studies. Theory that destabilizes the category of women, racial formations that disrupt the unity or primacy of the category, and sexualities that similarly blur the solidarity of the category – each of these must be resisted, restricted, or worse, colonized, to preserve the realm. (Brown, 1997: 83)

That feminist theory dwells on issues of identity is understandable. I need not revisit the now well-trodden history of theory's "end of inno-cence" (Flax, 1990).

Current concern with the fragmentation of identities is crucially linked to questions concerning the continued viability of differentiating between "sex" and "gender." Beginning during the Enlightenment, but only completely established in the 1950s, the "sex"/ "gender" binary has circulated throughout the social sciences, providing a powerful foundation for a material account of women's oppression. This bifurca-tion served a number of functions, most immediate of which was to provide a convenient, tangible means to constitute identity and proceed with the immediate concern of challenging the hierarchical relationships that subordinate women to men.

Confidence in this distinction is eroding, or has already degenerated to such an extent that Hood-Williams (1996) is able to offer its "post-mortem." Many feminist scholars have contributed to this "post-mortem" by critiquing the "sex"/ "gender" distinction.[14] For instance, Delphy argues that rather than seeing sex as the baseline from which gender emerges through sociality, "gender ... create(s) anatomical sex" (1984: 144). By conflating the biological with the natural, "sex" becomes the natural that initiates the social. Moreover, "natural" difference is almost entirely based on one particular aspect of biology: sexual reproduction, a view that will be strongly critiqued in Chapter 5. Under the discursive sign of sexual reproduction, an entire orchestra of "biological facts" are brought into play to fix the notion of biological "sex" differences. Thus, chromosomes, hormones, and genitalia have been variously "constituted as embodying the *essence* of sex" (Harding, 1996: 99 emphasis original).

Also critiquing the "sex"/ "gender" distinction, Hood-Williams (1996) focuses on three interrelated assumptions which underlie this (often) taken-for-granted binary. First, the biological distinction between women and men assumes that a distinction can be made between biology ("sex") on the one hand, and culture ("gender") on the other; and further, that while "gender" is changeable, "sex" is immutable. Finally, and most importantly for this book, this binary depends upon the idea that biology itself consistently distinguishes between females and males. Nature, as I argue throughout this book, offers shades of difference and similarity much more often than clear opposites, and that it is rather a modern ideology which imposes the current template of "sexual difference."[15]

Despite feminist critiques, the overall feminist project has largely depended upon a "real," corporeal base on to which "gender operates as an act of cultural *inscription*" (Butler, 1990: 146 emphasis original). Wittig's (1993) theory of lesbian identity illustrates Butler's point. Wittig argues that lesbians' position within the "sex"/ "gender" binary is ambivalent: lesbians are contemporaneously "women" (as defined morphologically) and "not women" (as defined by heteronormativity). The political project determined to challenge the heteronormative definition of "woman" makes Wittig's analysis valuable. However, the analysis relies on an immutable notion of "sex" to argue the social construction of "gender" (Fuss, 1989). While Wittig goes to some length to discuss lesbian *social* identity, she also quite clearly considers lesbian membership initially on the basis of *morphology*. Nowhere does Wittig discuss the possibility of lesbians with penises. So implicitly lesbians are women and women are females and females are human beings with a particular morphological body.

Even some postmodern feminists seem at times unwilling, in the final instance, to give up a corporeal notion of the feminine. At the same time that Shildrick, for instance, is able to write of "posthumanism" and "identity as process," she states "I ... have no wish to fully abandon the concept of the feminine" and that while "boundaries are fluid and permeable, they [do not] cease altogether" (1996: 9–10).[16] Postfeminist and cyberfeminist analyses may focus on the body as "fragmented" and "chimerical," but for most feminists these discussions remain conceptual, remote from the everyday material relations of "gender," where "sex" is fully grounded. As Paola Melchiori argues, "even 'gender' which was meant to escape nature traps, is ... becoming as rigid as nature in its exploratory capacity" (in Marshall, 2000: 43).

Within our current discursive field, to exist at all means being a woman or a man, or in Butlerian terms, "sex is the norm by which the 'one' becomes viable at all" (1993: 2). Thus, feminist theory continues to labor definitional concerns based on the "sex"/"gender" template. Indeed, on a practical level, much feminist theory continues to operate from a largely undisturbed two-sex model, as it appears to facilitate analyses of women's experiences.[17]

The challenge of this book is to explore how feminist theory might proceed from this point. In other words, the book takes as its problematic the challenge faced by feminist theory to look critically at the ways in which society, through culture and science, has structured a concept of "sex" that emphasizes difference rather than similarity. In the next chapter, I will examine the major bodily signs, which in contemporary Western society, are most often called upon to mark the physical "sexed" body: skeletons, hormones, chromosomes, and genes. But before we go on to look at these signs of contemporary notions of "sexual difference", it is important to signpost an essential ingredient of sexual difference that remains largely hidden from public scrutiny.

The missing link – heteronormativity

Thus far, I have argued that the contemporary emphasis on "sexual difference," refracted through the sex/gender binary, was established through epistemological and political shifts which took place over many years, and was particularly accelerated by the Enlightenment. However, this analysis is incomplete without the acknowledgment of heteronormativity as the undergird of the transition to a social order based upon sexual difference and complementarity. *Heteronormativity*, the hegemonic discursive and nondiscursive normative idealization of heterosexuality, played a leading role in establishing and then maintaining sex complementarity. The term "heterosexuality" was coined in 1892, as part of the growing interest of Victorian sexual science in surveying, labeling, and eventually treating an entire landscape of "perversions" including necrophilia, bestiality, and homosexuality (Ward, 1987). Today's normative idealization of heterosexuality took some time to establish and this history is a subject in its own right (see Foucault, 1979). The salient point here is the recognition of the dependence of the contemporary concepts of "sex" and "gender" on heteronormativity. As Judith Butler observes:

> 'gender' can achieve stability and coherence only in the context of a 'heterosexual matrix': a hegemonic discursive/epistemic model of gender intelligibility that assumes that for bodies to cohere and make sense there must be a stable sex expressed through a stable gender ... that is oppositionally and hierarchically defined through the compulsory practice of heterosexuality. (1990: 151, n. 6)

Butler is describing the dependence of the contemporary concepts of "sex" and "gender" on heterosexuality. That is, "sex" is defined as the biological *differences* between women and men, and these differences are defined in terms of their "function" (heterosexual procreation), "because of the presumption ... that the nature of heterosexuality is 'nature' itself" (Ward, 1987: 146).

The connection between the establishment of "sexual difference" and heteronormativity is important because it also fixes the modern use of "nature" as its foundation. For example, contemporary discussions of the importance of "the family" most often base their critique on the assumed interconnections between sexual difference, complementarity, and heteronormativity – females and males as "opposites" (sexual difference) who together (complementarity) form the most basic useful structure in society (heteronormative "purpose" of sexual reproduction). Critiquing what they see as "political correctness" gone crazy in Canada, Tom Darby and Peter Emberley write:

> Should the redressing of 'historical wrongs' be permitted to run roughshod over equality of opportunity? Should all representative political institutions correspond exactly to the statistical profiles of the population at large? Should the experiences of gender, race,

sexual orientation, and cultural difference be permitted to re-arrange *our* notions of the family, marriage, political decision-making or education? (1996: 239)

Barbara Marshall distills this form of critique by making the connections between sexual difference, complementarity, and heternormativity explicit: "Once we loosen the link between sex and gender, hell is but a short handcart ride away – the family will lie in tatters, one would be able to change one's sexual identity at will, the population will fail to replace itself ('obviously, more homosexuality, more women working outside the home, and less women seeing motherhood as natural would decrease the population')" (O'Leary in Marshall 2000: 103–4). Thus, "the family" has become a stand-in for heterosexuality, "that which is natural, desirable, and defensible as an ideal, and the ultimate location of immutable [sex] differences" (Marshall, 2000: 122). The concept of heteronormativity, as the largely unacknowledged specter of sex difference discourse, permeates the remainder of the analysis in this book.

Suggested readings

Laqueur, T. (1990) *Making Sex*. Cambridge, MA: Harvard University Press.
Schiebinger, L. (1993) *Nature's Body*. London: Pandora.

3
The Body of Sexual Difference

If I lived in a world with no racism or sexism, and where Catholics weren't urged to 'love the sinner and hate the sin' I might find more compelling the idea that people who are commonly recognized as 'born that way' are treated better. If I lived in a world where intersex infants were revered as gifted individuals who remind us of the natural multiplicity of physical sex, I might be happy to allow biologists to define what counts as the 'truth' about sexed bodies. If I lived in a world where you could produce easy journalism about the scandal of our kids getting bullied in school (perhaps in the Daily Mail?) as easily as you could about our finger-lengths, I'd be excited about the press coverage. However, I live in this world, so I'm not too jazzed. The issue is not nature vs. nurture. The issue is that we are offered new polished up degeneracy theories as an improvement over irredeemable sin. The version of nature that we are offered always positions us as deviant bodies, hyper-something, or 'lacking' in something else (a brain part, a hormone, whatever all the new essence of masculinity or femininity is supposed to be in this decade). The bodies that supposedly produce heterosexual people in these theories are always treated as normative. I'll get excited about this stuff when heterosexual people start to do the epistemological work of worrying about how they could have been born that way, or whether it was their parents 'fault'. Until then (as Freud, for all his faults, recognized) we're not dealing with an epistemology that can recognize difference (biological or otherwise) without placing people in a hierarchy – and theories like that are not going to go anywhere, politically, or scientifically.

(Hegarty, Lesbian, and Gay psychology
mailing list, October 28, 2003)

29

Introduction

Chapter 1 of this book described a growing body of feminist theory focused on the "culture of matter," or what Evelyn Fox Keller terms the "social construction of science" (1989: 34). A short list of such critiques includes (but is certainly not limited to) *Women Look at Biology Looking at Women* (Hubbard, Henifin, and Fried, 1979), *Genes and Gender* (Tobach and Rosoff, 1978), *Science and Gender* (Bleier, 1984), *Myths of Gender* (Fausto-Sterling, 1992), *Nature's Body* (Schiebinger, 1993), *Paradoxes of Gender* (Lorber, 1994), *The Century of the Gene* (Keller, 2000), and *Feminist Science Studies* (Mayberry, Subramaniam, and Weasel, eds 2001). This chapter focuses on critical feminist analyses of what might be termed the "essence" of "sexual difference." This "essence" consists of bone structure, gonads, hormones, chromosomes, and genes (the list also includes sexual reproduction, the critical analysis of which is so extensive as to merit its own chapter – see Chapter 5). I have derived this list mainly from several years of discussion with students about "sex," "sexual difference," and "sexuality," as well as attending to media accounts of sex "differences." As Chapter 1 detailed, however successful feminist arguments concerning the social construction of gender have been within academia and the public in general, there remains a persistent and robust recourse to a biological notion of "sexual difference" based upon often cursory notions of testosterone levels or X and Y chromosomes. For this reason, I want to employ feminist theory to critically review each of these "facts" of "sex" with a view to highlighting the mechanisms through which scientific knowledge is constructed.

Patricia Gowaty defines science as the "practice of systematic observation and experiment as a means to test predictions from hypotheses while reducing or eliminating (i.e., controlling) the effects of perceived and possible biases on results and conclusions" (1997b: 14). Feminist critics of scientific knowledge are mainly concerned with the *processes* of scientific research insofar as biases may be introduced which influence the outcome of research (for a fuller explanation see Hubbard 1979). The weak version of this critique is that the propensity for the introduction of bias into scientific research limits the degree to which scientific knowledge can claim objectivity. The strong version argues that insofar as science is based upon a set of knowledge claims, it is necessarily limited by the parameters of this knowledge. Or as Gowaty more succinctly argues, " 'objective knowledge' is an oxymoron" (1997b: 14). Let us take a recent example popularized in the media which illustrates the point that feminists make about the problem of objectivity in science studies of, in this case, sexuality.

As I will argue in Chapter 6, one of the most significant corollaries of the classification of homosexuality as a sexual practice has been the unending search for its "cause." Perhaps the most recent variation of this search is found in the work of neuroanatomist Simon LeVay (1991) who has spearheaded research searching for differences in brain structure between homosexual and heterosexual people. LeVay's research focused on the interstitial nuclei of the anterior hypothalamus (INAH), the region of the brain thought to be involved in basic life functions such as respiration, circulation, metabolism, and sexual behavior. This region of the brain has been classified into four parts: INAH1, INAH2, INAH3, and INAH4. LeVay's research concluded that while no differences in size could be discerned for INAH1, 2, and 4, there were significant differences in the size of INAH3 nuclei between homosexual and heterosexual men: homosexual men's INAH3 was almost three times smaller than heterosexual men's.

A number of critics of this research have pointed out the serious methodological flaws of LeVay's research. For instance, LeVay's sample was derived from 41 cadavers: 19 presumably homosexual men, 16 presumably heterosexual men, and 6 presumably heterosexual women. LeVay makes the assumption that 19 of the male cadavers were homosexual because all died with AIDS, presumes the 16 men were heterosexual because although they all died with AIDS they were intravenous drug users, and presumes the remaining 6 female cadavers were heterosexual although he admits the women's sexual orientation was not noted in their medical files. LeVay defends his study against criticisms that he presumed the sexual behavior of his sample of cadavers, that the AIDS virus might have affected the results, that measuring the INAH of only men infected with AIDS produced an unrepresentative sample, and that the sample size was much too small to be representative in any case (Murphy, 1997). Moreover, LeVay chooses to highlight supposed differences between presumably heterosexual and homosexual men's brains, rather than the intravariability of the findings. For instance, while the overall findings may have suggested that gay men have smaller INAH3 nuclei, the second-largest INAH3 belonged to a gay man and the third smallest INAH3 belonged to a heterosexual man (Murphy, 1997). That is, the standard deviation of variability was considerable, with some gay men and some heterosexual men having smaller INAH3 nuclei and other gay men and heterosexual men having larger INAH3 nuclei.

Beyond these significant methodological problems with the study, there are a number of larger problems that underscore the stronger version of the argument that scientific knowledge claims are epistemologically limited by the social construction of what "counts" as scientific knowledge.

In the popularized *The Sexual Brain* (1993), LeVay begins by outlining the arguments for exploring the biological basis of sex differences. He argues that many people overestimate environmental influences on individual "sex-typical" behavior. But throughout the book, LeVay himself is dependent upon cultural conceptions of sexual dimorphism that he himself appears to acknowledge are socially constructed. First of all, the bulk of the research cited throughout the book is on rats, not humans. Yet, discussing the effects of prenatal androgens on sexual games that children apparently play, LeVay states "this claim seems to depend upon a false assumption about the equivalence of hormonal mechanisms in rats and humans" (1993: 89). He goes on to submit that there is considerable variation in nonhuman animal sexual behavior and that this variation is at least partly influenced by contact with playmates – that is, environmental influence. LeVay also makes the typical cultural distinctions necessary for emphasizing sex differences. Thus, male rats engage in "rough-and-tumble" play whereas female aggressiveness is defined as 'maternal aggression' and thus does not seem to count as actual aggression, or is at least limited to the role of mother. Moreover, although LeVay admits that "there are many sexual practices that do not involve distinct 'masculine' or 'feminine' roles," these are not discussed with respect to nonhuman animals (1993: 106). Only those behaviors that emphasize sex "differences" seem to count in the discussion. For example, LeVay thinks that

> most people would agree that there are subsets of gays and lesbians who are markedly sex-atypical in behavior, skills and interests who gravitate to occupations where these traits are of value. Jobs requiring leadership and organizational skills; mechanical, electrical, and transportation jobs; athletic and strongly physical occupations; all seem to be especially attractive to a subset of lesbian women. Jobs requiring creative and caring traits – design, writing, dance, theatre, nursing, and so on – seem especially attractive to some gay men. (1993: 119)

Notice this opinion is based upon "seem," "some," and cultural definitions of what is "sex-typical" and "sex-atypical." He even goes on to say that "on the whole it appears that sex-atypical traits are more uniformly seen in children destined to become gay or lesbian than in adults who *are* gay or lesbian." How has LeVay determined that the children he refers to do actually become lesbian or gay? Clearly some of LeVay's arguments are based upon cultural interpretations, such as his understanding of differences in lesbian and heterosexual women's occupational status. Perhaps the social construction of what is considered scientific knowledge is most transparent when LeVay states that "I would further guess on the basis of my own discussions with lesbian women" (1993: 115).

It is difficult to argue for the "objectivity" of scientific knowledge when LeVay's suppositions are based upon his personal discussions with lesbian women he knows. In another example of LeVay's propensity to socially construct scientific knowledge, he regrets that "unfortunately, these ideas are too nebulous to test in any rigorous fashion" (1993: 103). Anne Fausto-Sterling makes the important point that LeVay's study is based upon a strongly dimorphic classification of "heterosexual" and "homosexual" behavior such that his sample is presumed to fit easily into these supposedly exclusive categories (in Murphy, 1997). Fausto-Sterling persuasively argues that the complexity of sexual behavior precludes such facile classification (a point taken up in Chapter 6). Part of the problem is the continuous confounding of concepts such as "sex" "sexuality" and "sexual difference" (see Chapter 1). For instance, certain behaviors found in nonhuman animals (more specifically, rats) are assumed to be sexually dimorphic and thereby evidence of sexuality or sexual orientation. But as Feder argues, these behaviors have nothing to do with sexual orientation, and to assume so is to confuse the questions: "who is a person sexually attracted to?" and "what role does a person take when he or she has sex?" (in LeVay, 1996: 119). Ruth Hubbard and Elijah Wald (1993) also criticize LeVay's study by pointing out that the reason scientific studies continue to focus on the "cause(s)" of homosexuality is because homosexuality is stigmatized in society. Rather than focus on homosexuality *per se* with an either explicit or implicit view to eradicating or "alleviating" homosexuality, why do studies not focus on the "causes" of the stigmatization, with a view to eradicating this form of prejudice and discrimination? Celia Kitzinger asks an important question of biological studies of lesbianism and homosexuality: "… what are we trying to prove with all these studies claiming a biological basis to homosexuality? It may or may not be biological, but what's that got to do with whether or not it's okay for people to beat us up outside gay pubs, insult our children in playgrounds and discriminate against us in law and social policy" (email correspondence on Lesbian and Gay psychology mailing list, October 28, 2003).

To reiterate, feminist studies of the social construction of science argue, *at a minimum,* that biases limit the degree of objectivity possible in science. The strong version argues there can be no objectivity in science period. Throughout the book I will re-visit the tension between these two approaches to science.

The "essence" of sexual difference

The remainder of this chapter explores what are typically cited as the "facts" or "essence" of sex differences. In this exploration I focus on how feminist scholars argue that these "facts" are mediated by the social construction of scientific knowledge: a social construction based upon the "two-sex" model that seeks to emphasize sex differences rather than similarities.

Skeletons

Recall from Chapter 2 that the Enlightenment brought about a slow yet eventually persistent shift from a "one-sex" model in which females and males shared one sex to the current "two-sex" model of sexual dimorphism whereby females and males are understood to have completely separate morphologies. The search for morphological differences between females and males was thorough, initially focusing on the exterior of the body, for instance in analyses that emphasized the penis and clitoris as completely separate morphological structures (indeed, the clitoris was entirely usurped by the vagina as part of the emphasis on sex "differences"), and extending to the interior of the body as medical techniques such as the autopsy were refined. The focus on human skeletal structures provides a good example of the search for morphological sex differences. The skeletons that Vesalius, the most prominent anatomist of the 1500s, drew were simply labeled "human skeleton" underlining that whatever sex "differences" might appear on the surface of the body (for instance, breasts) were only skin deep, and did not extend to "deep" anatomical structures such as skeletons (Schiebeinger, 1993). By the late eighteenth century, anatomists began to provide evidence for the notion of "sex" complementarity by drawing distinct female and male skeletons.

For instance, in 1765, the French *Encyclopdie* produced a direct comparative analysis of female and male skeletons, arguing that differences in the skull, spine, clavicle, sternum, coccyx, and pelvis proved that "the destiny of woman is to have children and to nourish them" (in Schiebinger, 1993: 222). At the time, the skeleton was viewed as the most penetrating and "deep" aspect of the human body: differences between females and males at the center of the body could only mean differences throughout the rest of the body, in the muscles, organs, and veins.

Of course, the differences that anatomists and doctors found between female and male skeletons were given meaning. Londa Schiebinger notes that the female skeleton became the signifier of not only a completely different physical and mental constitution, but also a different purpose in life: "1) a weak constitution makes the bones of women smaller in proportion to their length than those of men; 2) a sedentary life makes their clavicles less crooked (their arms are hindered by their clothing and have been less forcibly pulled forward); 3) and a frame proper for their procreative functions makes women's pelvic area larger and stronger to lodge and nourish their tender fetus" (1993: 157). We find a similar conditioning of women's constitution in the debate between Thiroux d'Arconville and Soemmerring concerning the size of women's skulls. Whereas d'Arconville argued that women's skulls were smaller than men's, Soemmerring argued that women's skulls were actually larger than men's. And yet, this finding does not lead Soemmerring to conclude anything about women's capacity for thinking or rationality, but actually *confirms* women's inferiority to men: "women lead a sedentary life and consequently do not develop large bones, muscles, blood vessels and nerves as do men; since brain size increases as muscle size decreases, it is not surprising that women are more adept than men in intellectual pursuits" (in Schiebinger, 1993: 207). Indeed, women's larger skulls became evidence of their incomplete development, and similarities were drawn to children who also have larger skulls relative to their bodies.

Interestingly, the human bodies on which these skeletal drawings were based were not themselves sex dimorphic. In other words, anatomists used several different bodies, sometimes male and sometimes female, as composites for the "typical" female and "typical" male skeleton.[18] Thus, the female skeleton produced for a given book on sex complementarity might actually be comprised of one woman's pelvic bones that matched the cultural ideal of suitability for childbearing, added to another woman's skull that met the cultural ideal of female irrationality, added to yet another woman's ribs that exemplified the cultural ideal of a narrow and fragile chest. This type of composite drawing necessarily downplayed intrasex differences, that is difference within each sex, as it literally "picked and chose" skeletons that matched cultural ideals of sex complementarity. As Schiebinger argues, "anatomists in the 18th century 'mended' nature to fit emerging ideals of masculinity and femininity" (1993: 203). More recently, Alan Peterson (1998) has analyzed shifts between 1858 and the present in medical representations of female and male skeletons in Gray's *Anatomy*. Peterson notes that increased emphasis on comparisons between the two skeletal structures concomitant with an emphasis on the superiority of the male body, serves to emphasize comparatively miniscule sex differences while minimizing much more obvious similarities.

Gametes

Gametes are defined as sex cells. These include sperm and eggs in human animals. In human animals, gametes develop and are stored in the gonads; sperm in the testicles and eggs in the ovaries. A number of feminist scholars have critiqued the processes through which gametes have historically come to reinforce a highly stereotyped understanding of sex "differences." For instance, Ruth Hubbard (1979) argues that much of sociobiology is based upon an exegesis of Darwin's original notion of sexual selection. Although this will be discussed more extensively in the following chapter, in its most androcentric form, sexual selection posits that females of most species are more heavily invested than males in the conception and care of offspring because of their disproportionate investment of energy in their offspring. Females are not only assumed to be disproportionately responsible for the care of live infants, but the theory extends to the gametic level, where the differences in size, and numbers produced, of eggs and sperm also seem to account for sex differences in parental investment. George Williams, in *Sex and Evolution* (1975) puts it this way: "the essential difference between the sexes is that females produce large immobile gametes and males produce small mobile ones" (in Hubbard, 1979: 24). The supposed differential energy investment by females and males is accounted for by the fact that females produce fewer eggs than males produce sperm, and by the characterization of eggs as passive and immotile compared with the greater activity and motility of sperm. However, Hubbard (see also Snowdon, 1997) queries this assumption of energy investment by pointing out that the average woman produces about four-hundred eggs in her lifetime, compared with the several billion sperm produced by the average male. On average, in contemporary Western societies, women and men will invest in 2.2 (or less) live offspring. This means that the female ratio of investment is only about 400:2 compared with the male investment of several billion : 2. Hubbard makes the point that no one actually knows the comparable energy investment in producing eggs and sperm, and yet this does nothing to dissuade androcentric theories of greater female investment. In "The Egg and the Sperm: How Science Has Constructed a Romance Based on Stereotypical Male–Female Roles" (1991), Emily Martin further argues that the language in which gametes are discussed in the scientific literature belies an androcentric bias toward qualities associated with masculinity. Martin notes that even though, compared with egg production, sperm are extremely disproportionately wasted (i.e. do not result in offspring), sperm production is never discussed in terms of "waste."

Martin further details the socially constructed processes of scientific research on gametes. Martin's central argument is that the cultural valuation of masculinity over femininity produces a highly distorted presentation of egg and sperm composition, activity, and function within cell biology. Interestingly, Martin identifies the mechanism through which gametes are socially constructed to be the principle of equality. That is, Martin argues that egg and sperm are anything but equal in terms of size, activity, or function. Eggs are much larger than sperm, eggs contribute the entire cytoplasm (containing nutrients and mitochondria which contains DNA) as well as messenger RNA, most of the nucleoprotein complexes that provide much of the proteins for fertilized eggs, ribosomes, and the cell membrane (Figure 3.1) (Figure b shows the size of a spermatozoa in relation to an egg).

Emphasizing the equality of egg and sperm constructs a scientific narrative that renders invisible much of the greater activity and function of the egg. Part of this mechanism involves an emphasis on the importance of the nucleus as opposed to the cytoplasm and does not take into account that the mitochondrial DNA contributed entirely by the cytoplasm means that females actually contribute more than 50 percent of the DNA to offspring than do males. Indeed, although gametes have long been the subject of sustained scientific research, only lately has the crucial contribution of the cytoplasm in the form of nutrients, the structures essential for activity and DNA, been studied at all.

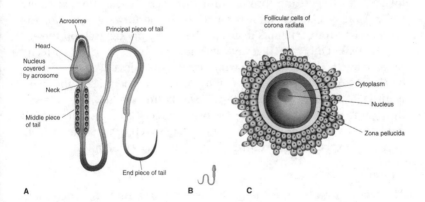

Figure 3.1 Illustration of a human egg and spermatozoa

Source: Moore, K. and Persaud T. (1998) *Before We are Born: Essentials of Embryology and Birth Defects*, 5th edition. Philadelphia, PA: W.B. Saunders Company, p. 18. Reprinted with permission from Elsevier.

Martin also details the ways in which stereotypical understandings of femininity and masculinity are imported into discussions of gametes as yet a further mechanism of the socially constructed narrative within scientific literature:

> it is remarkable how 'femininely' the egg behaves and how 'masculinely' the sperm. The egg is seen as large and passive. It does not *move* or *journey*, but passively 'is transported', 'is swept' or even 'drifts' along the fallopian tube. In utter contrast, sperm are small, 'streamlined', and invariably active. They 'deliver' their genes to the egg, 'activate the developmental program of the egg' and have a 'velocity' that is often remarked upon. Their tails are 'strong' and efficiently powered. Together with the forces of ejaculation, they can 'propel the semen into the deepest recesses of the vagina'. For this they need 'energy', 'fuel', so that with a 'whiplashlike motion and strong lurches' they can 'burrow through the egg coat' and 'penetrate' it. (1991: 489)

Martin notes that the scientific literature tends to emphasize the fragility of the egg, both in terms of its supposedly limited motility (yet the egg travels down the fallopian tube unaided by sperm) and its inability to survive on its own once released, while deemphasizing the fact that sperm also die within a few hours of release. Interestingly, recent research on sperm activity has found that contrary to the image of sperm as forceful seekers of eggs, the forward thrust of sperm is actually extremely weak. Moreover, much more strongly than the forward thrust of the sperm is its strong attempts to escape the egg by prying itself off the egg (1991: 493). Only the digestive enzymes of the sperm, if correctly released by the tip of the sperm while leaving the sides of the sperm stuck on the egg, enables the "weak, flailing sperm" to orient itself and make it through the egg cell walls (1991:493). Martin concludes her analysis by arguing that "the implanting of social imagery on representations of nature ... lay[s] a firm basis for re-importing exactly that same imagery as natural explanations of social phenomena" (1991: 500).

Hormones

Theories concerning the relative importance of hormones in determining sexed behavior are, of course, dependent upon the invention of medical and biological technologies to isolate particular families of molecules within the blood of first nonhuman, and then human animals. Before the invention of these techniques, the locus of sex dimorphism was located, as Chapter 2 detailed, in notions of heat and humors, followed by particular organs such as the penis and uterus. Until the twentieth century, women did not refer to "sex hormones" to describe events in their lives (such as feelings after childbirth), because this term did not exist. It was not until the beginning of the twentieth century that the study of hormones and sex differences developed into what was eventually termed "sex endocrinology." This field of science built upon geneticists' notion that sex was determined at birth (by genetic factors), by arguing that this genetic endowment was augmented by both physiological and environmental conditions during embryonic development. Nellie Oudshoorn (1994) maintains that the study of "sex hormones" was enabled by a particular relationship between laboratory scientists, clinicians, and pharmaceutical entrepreneurs who focused on either biological or chemical aspects of the body.

Initial studies of "sex hormones" posited a simple and direct relationship between "sex" and hormones: females had a female sex hormone that originated in the ovary and males had a male sex hormone that originated in the testes. Thus, each "sex" was understood to be designated one sex hormone, and moreover, that female and male sex hormones were antagonistic to each other. Needless to say, this dichotomy suited the idea of sex complementarity, and it should not be surprising that this dichotomy was slow to be challenged. However, it was eventually discovered that "sex hormones" were not limited to human animals. First, plants and fungi were found to be rich in "female" hormones. Next, "female" hormones were found in male animals, as this article published in 1934 concedes:

> Curiously enough, as a result of further investigation, it appears that in the urine of the stallion also, very large quantities of oestrogenic hormone are eliminated ... I found this mass excretion of hormone only in the male and not in the female horse. The determination of the hormone content, therefore, makes harmonic recognition of sex possible in the urine of the horse. In this connexion we find the paradox that the male sex is recognized by a high eostrogenic hormone content. (Zondek in Oudshoorn, 1994: 25–6)

Eventually, "female" sex hormones were found in human males. Yet, the idea of sex hormone exclusiveness persisted and scientists suggested various theories to explain the appearance of hormones in the "opposite" sex, including latent intersexuality and that the ingestion of certain foods produced levels of "opposite" sex hormones.

In time, the theory of one female sex hormone and one male sex hormone was abandoned to a new theory that "sex hormones" existed in various groups. The first hormone to be purified was estrone from the *estrogen* family in 1929, next came progesterone from the *progestin* family in 1934, and then testosterone from the *androgen* family the next year (LeVay, 1996). The classification of different families of "sex hormones" led to the theory that the "female" sex hormones in males were ineffective as were the "male" sex hormones in females. Other theories supported the idea that some families of "sex hormones" were "more" male than female (testosterone) while other families of hormones were "more" female than male (estrogen) and still other families of hormones were "bisexual" (androgen). The very close resemblance between these families of chemical compounds (differing by just one hydroxyl group – see Figure 3.2), led scientists to conclude that "there but for one hydroxyl group go I" (in Oudshoorn, 1994: 39).

Another popular theory during this period, and to some extent today, was that "opposite" sex hormones are correlated with homosexuality. Hormones have been extensively studied in relation to homosexual behavior in nonhuman animals. Indeed, from the 1930s onward, "male" hormones have been injected into females and vice versa to see changes this might produce in sexual behavior (Dagg, 1984). Organotherapy is the attempted "conversion" of homosexual people to heterosexuality through hormone injections. From the 1930s to the 1980s, numerous studies attempted to demonstrate a relationship between hormones and sexual orientation. Twenty studies found no differences between testosterone levels in gay and heterosexual men; two studies found the testosterone levels of gay men to be *higher* than that of heterosexual men; and three studies found the opposite (LeVay, 1996).

The purported association between hormones and homosexuality has been widely studied. In the 1930s, homosexual men were treated with large doses of hormones in an attempt to produce heterosexual behavior. Contemporary studies mainly focus on hormones such as testosterone and gonadotropins (Luteinizing Hormone and Follicle Stimulating Hormone). As we know, testosterone is most often labeled a "male" hormone, even though it is found in both males and females. An erroneous understanding of homosexuality in males as "feminine" prompts some researchers to hypothesize that gay men will have lower levels of testosterone than heterosexual men. A similarly erroneous understanding of lesbianism as "masculine" has led some researchers to hypothesize that lesbian women will have lower levels of Luteinizing Hormone and Follicle Stimulating Hormone. In reviewing these studies, Nanette Gartrell (1982) notes these studies fail to find any positive association between levels of hormones and homosexual behavior. Indeed, the

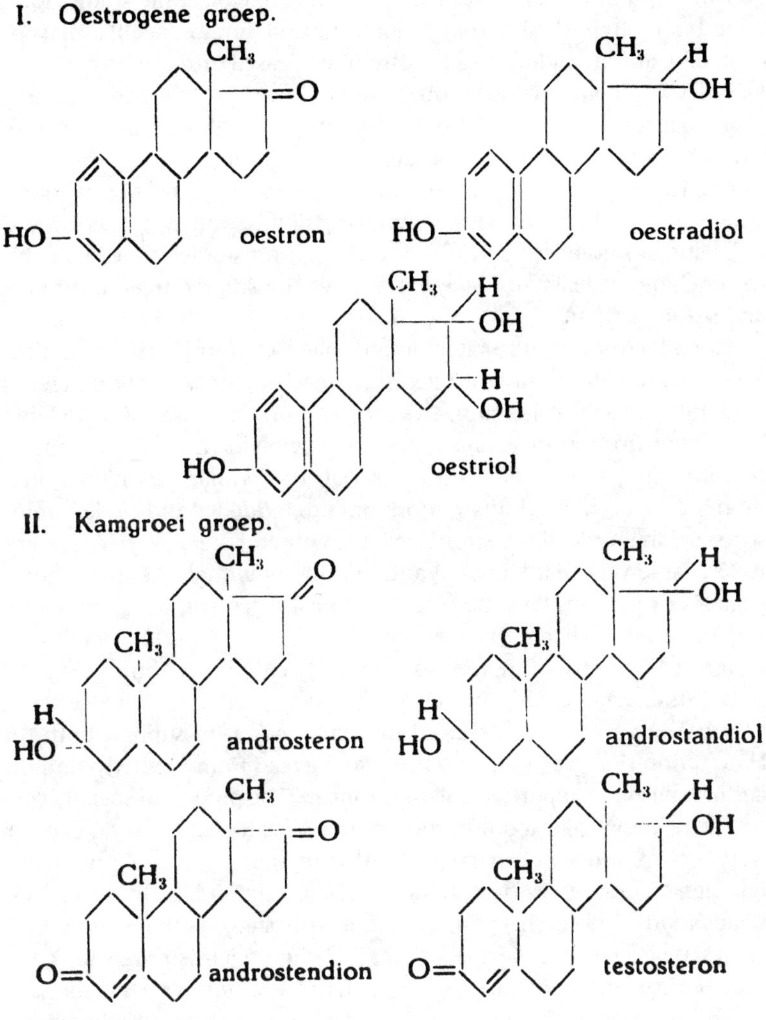
I. Oestrogene groep.
CH3
=O
HO–
oestron
CH3
H
OH
HO–
oestradiol
CH3
H
OH
H
OH
HO–
oestriol
II. Kamgroei groep.
CH3
O
CH3
H
HO
androsteron
CH3
H
OH
CH3
H
HO
androstandiol
CH3
=O
CH3
O=
androstendion
CH3
H
OH
CH3
O=
testosteron

Figure 3.2 Early illustration of hormone compounds

Source: Oudshoorn, N. (1994) *Beyond the Natural Body: An Archeology of Sex Hormones*. London: Routledge, p. 30. Reprinted by permission of Taylor and Francis. Originally appeared in: Freud, J. (1936) "Over Geslachtshormonen," Chemisch Weekblad 33, 1. Reprinted by permission of Chemisch Weekblad.

studies as a whole suggest that males exhibiting heterosexual behaviors and males exhibiting homosexual behaviors have comparable total testosterone and gonadotropin measurements. The few studies that focus on lesbian women concluded similarly that testosterone and gonadotropin levels did not predict sexual behavior. Some studies have gone on to hypothesize that lesbianism and homosexuality may be caused by abnormal levels of testosterone released into the body during fetal development. However, these studies have also failed to show any positive association. Gartrell notes that it is difficult to draw firm conclusions from these studies because any hormone resistance or deficiency in the fetus can only be inferred from postnatal observations. Finally, studies abound that attempt to attribute some sort of hormonal imbalance to homosexual activity, although no studies have been able to conclude the existence of any causal relationship between hormones and sexual activity.

Sex endocrinology ultimately found that "sex hormones" are not sex exclusive. It also found that "sex hormones" are only finely differentiated (by one hydroxyl group), and that sex hormones are converted into each other (for instance, *aromatase*, an enzyme in the brain, is capable of converting testosterone into *estradiol*, the primary estrogenic hormone). However, the ability of hormones to influence and/or determine sexed behavior remains a strong belief in contemporary Western society. In the same way that Hubbard and Martin argue that scientific knowledge about gametes have been socially constructed within an androcentric paradigm that emphasizes both sex dimorphism and traditional conceptualizations of femininity and masculinity, Jennifer Harding (1996) and Nellie Oudshoorn (1990, 1994) argue that knowledge about "sex hormones" was constructed in such a way as to support both the assumption that "sexual difference" can be read from the body, and the cultural need to support sexual dimorphism. The process of socially constructing knowledge about hormones to coincide with cultural conceptions of sex dimorphism begins with the very naming of particular families of molecules "sex steroids" or "female and male hormones" (Oudshoorn, 1994). This naming erroneously suggests that certain molecules are exclusive to females or males, when scientists are well aware that both women and men have "female" hormones (estrogens) and "male" hormones (androgens). Moreover, hormones are interconverted in the body, and their relative proportions change throughout the life cycle such that post menopausal women have, on average, lower levels of estrogen and progestin than men of the same age (Oudshoorn, 1994).

Celia Roberts (2003b) makes the point that in the twenty-first century hormones are no longer considered to be contained within internal systems. Research on the interactions between hormones found in the environment and bodies shows that "contemporary hormones and chemicals acting like hormones in the environment do not respect boundaries: of space (between countries or bodies), of species, or indeed of time ... Does it make sense in today's hormonal world to think of either human or non-human bodies as discrete entities?" (2003b: 7. See Chapter 5 for more on hormones).

Genes

In the nature/nurture joust for the "essence" of sexual difference, the only topic to supercede hormones in the popular imagination is genes. A *gene* is a molecule of DNA. Put another way, genes are composed of DNA structures; DNA stands for deoxyribonucleic acid, which form long chain molecules. Since their discovery in 1953, genes have increasingly taken center stage in arguments for and against the "nature" of sexual difference, and the "nature" of just about everything human for that matter. The field of genetic research is changing rapidly, and each new advance in understanding brings a more complex picture of both the structure and function of genes.

The discovery of genes is usually credited to the Austrian monk Gregor Mendel, whose cross-pollination of pea plants led him to suppose the existence of some sort of hereditary factor in breeding. Later, William Bateson coined the term "genetics" to describe the branch of science devoted to investigating the biology of inheritance. But it was in 1953 that James Watson and Francis Crick first identified the now entirely familiar double helix pattern that make up the strings of DNA molecules (Figure 3.3).

From that time, the science of genetics burgeoned. Physics had the atom, chemistry the molecule, and genetics soon identified the gene as the fundamental unit of explanation. In 1990, the Human Genome Project (HGP) was launched with the express intent of identifying the exact sequence of genes in the human body. As a first step, the genetic sequence was identified in the bacterium *Escherichia coli*, then the roundworm *Caenorhabditis elegans*, and then the fruit fly *Drosophilia*. Geneticists have now sequenced the DNA structure for the human body, with some interesting results. I say "interesting" here because the results of genetic research have thus far presented as many challenges to, as reifications of, scientific and popular assumptions about the determinist properties of genes. For instance, at the inception of the HGP, many scientists claimed that knowing the genetic sequence would provide all the necessary information about biological function (Keller, 2000). Now that several animals' DNA has been sequenced, scientists realize that the

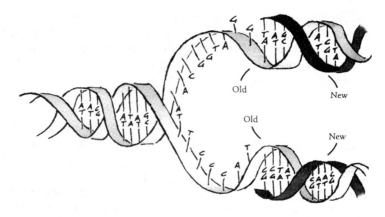

Figure 3.3   Illustration of a gene
Source: Reprinted by permission of the publisher from *The Century of the Gene* by Evelyn Fox Keller, p. 24, Cambridge, Mass.: Harvard University Press, Copyright ©2000 by the President and Fellows of Harvard College.

sequence or structure of the gene does not hold the explanatory power that was initially supposed. Most importantly, perhaps, is the idea that genes are tools that work in tandem with a number of other molecules at the biological level. But more on this later. Let us consider first of all what we know so far about genes.

In terms of our DNA structure, human beings have approximately 3 billion base pairs, while corn and salamanders have more than 30 times that number. By conservative estimates, human animals share about 98.5 percent of their genetic structure with chimpanzees (humans have 46 chromosomes; chimpanzees have 48 chromosomes). Eighty-five percent of human genetic variation occurs within any nation, and only fifteen percent of genetic variation can be traced between nations within any given race, and between races (Margulis and Sagan, 1997). Moreover, all Europeans are thought to be a hybrid population with 65 percent Asian and 35 percent African genes (Keller, 2000). At the chromosomal level, while no two people (except identical twins) have the same chromosomal constitution, all humans share 99 percent of their chromosomes. In other words, the differences which we hold so dear (hair color, skin tone, etc.) and on which so much of our social organization is based ("sex" segregated sports is an obvious example) are minuscule in comparison with our biological similarities.[19] About 90 percent of human DNA has no known function and is referred to as 'junk' (Rabinow, 1992). Only the remaining 10 percent of our DNA is transcribed into RNA and then coded for proteins.[20]

So far, then, investigations into genes have not led to any robust declarations of either the uniqueness of human animals from other animal species, or from each other. Part of the problem here is that genes have been acceded too much explanatory power. At the inception of the HGP scientists theorized that genes were the ultimate structure which controlled the production of bodily processes, and ultimately behavior. For instance, Evelyn Fox Keller paraphrases this theory as the "central dogma" of scientific research; that "DNA makes RNA, RNA make protein, and proteins make us" (2000: 54). Moreover, the public seems to believe that genes are somehow constitutive of an organism, that genes are essential traits rather than:

> sites of biological processes with variable outcomes, that they are activators rather than acted on, and that in some ways they are miniature people waiting to emerge in full somatic form and behavior. Genes are read as the molecular equivalent of manifest human traits – as little human beings preformed in the ribbons of DNA and merely awaiting fleshy instantiation in order to unfold according to a predetermined logic'. (Murphy, 1997: 180)

Thus, Gelbart argues that above any physical entity, genes are more accurately defined as "concepts that have acquired a great deal of historical baggage over the past decades" (1998: 660).

However, behind this public perception, the more scientists study genes, the more it becomes clear that there is no simple relationship between the sequence of DNA base pairs and the functional activity of proteins they are supposed to code for. As Gray argues, "although the nucleotide sequence does specify the primary structure of a protein (its sequence of amino acids), it is the tertiary structure of the protein that determines its function, and this depends on a range of nongenetic chemical and physiological factors inside the cell" (Gray, 1997: 389).

First of all, only about 3 percent of human genes code for proteins because many genes regulate rather than structure. Second, there is no one-to-one relationship between genes and proteins: there are hundreds of proteins associated with some genes, fundamentally challenging the assumption that one gene produces one protein. Nor do genes themselves determine which proteins should be made under particular circumstances: this is determined by a much more complex process of the whole cell. Evelyn Fox Keller likens this highly complex cell process to a musical score:

> ... the problem is not only that the music inscribed in the score does not exist until it is played, but that the players rewrite the score (the mRNA transcript) in their very execution of it. Furthermore, such genes have none of the permanence traditionally expected of genes – these recompiled mRNA transcripts are called into being only as needed and generally have rather short lifetimes. Indeed, they do not reside on the chromosomes and, in some cases, might not even be found in the nucleus – that is, the final version of the transcript may be put together only after the original transcript has entered the cytoplasm. (2000: 63–4)

This emerging picture of genes is dramatically different to the media constructed picture that genes structure and control everything in the body. Genes are much more dependent upon cellular metabolism:

> In fact, left to its own devices, DNA cannot even copy itself: DNA replication will simply not proceed in the absence of the enzymes required to carry out the process. Moreover, DNA is not intrinsically stable: its integrity is maintained by a panoply of proteins involved in forestalling or repairing copying mistakes, spontaneous breakage, and other kinds of damage incurred in the process of replication. Without this elaborate system of monitoring, proofreading, and repair, replication might proceed, but it would proceed sloppily, accumulating far too many errors to be consistent with the observed stability of hereditary phenomena – current estimates are that one out of every hundred bases would be copied erroneously. With the help of this repair system, however, the frequency of mistakes is reduced to roughly one in 10 billion. (Keller, 2000: 26–7)

Genes cannot replicate themselves nor persist over generations. Because genes work in necessary tandem with the cell's cytoplasm, genes are not the immortal messengers of the "essence" of human beings passed from generation to generation. Ho, Saunders, and Fox note that "heredity" resembles less a linear chain of command and more an "inter-locking feedback" process between nucleus, cytoplasm, and cells. Inheritance is thus "a property of the whole system, not just the genes in the nucleus" (1986: 43). Similarly, in *The Ontogeny of Information* (2000), Susan Oyama convincingly argues that information itself has its own ontogeny in that developmental information develops from the *contingent* relationship between genes and the environment (Gray, 1997). Gray likens this argument to a literary text. Rows of letters have no intrinsic meaning, except in the context of a reader in a particular culture and experience. So too DNA sequences have no intrinsic meaning except in the context of its protein environment and the context of its "reading": Gray refers to this as the difference between a "distributed program and a genetic program" (1997: 146). Moreover, in more complex organisms, DNA sequences do not automatically translate into amino acid sequences, but are dependent upon the state of the cytoplasm they are in.

There is one type of DNA that is more interesting in relation to inheritance. The study of the DNA found in the mitochondria of each cell has been dubbed "The Other Human Genome Project" because this type of DNA has been, until recently, almost completely overlooked (Palca, 1990). Mitochondria are essential to any cell, as they provide the metabolism of the cell by taking energy from organic molecules and transforming it into adenosine triphosphate (ATP) which is used to power the cell.[21] Interestingly, the DNA in mitochondria is inherited entirely from the mother. So the public perception that children inherit equal genetic material from their mothers and fathers is false: all children are more genetically related to their mothers than their fathers. In other words, the majority of any human being's DNA is inherited matrilineally. Indeed, through mitochondrial DNA, biologists have been able to trace the first *Homo sapiens* to about 600 thousand years.

What about genes and "sex"? Human beings have 23 pairs of chromosomes, or 46 chromosomes in total. Forty-four of these chromosomes are not related to "sexual difference" in any way. But because our society focuses so much on "sexual difference", we tend to concentrate on the two chromosomes (or one pair) that are related to "sexual difference": denoted as the X and Y chromosomes for their appearance. This pair of chromosomes is usually defined as XX (homogametic) for females and XY (heterogametic) for males. However, there are in actuality many variations of "sex" in humans: XXY, XXXY, XXXXY, XXYY, XXXYY to name only a few. There is also great diversity in nonhuman animal chromosome structures: male birds are homogametic with two Z chromosomes and females are heterogametic with one Z and one W chromosome – thus female birds determine the sex of their offspring (Snowdon, 1997). Some reptile and amphibian species have no sex chromosomes, and the sex of offspring is determined by the temperature that the eggs are incubated at. Some fish species are either sequentially or simultaneously intersex. Therefore, chromosomes are far from a fail-safe means of determining sex. Moreover, the role played by social and environmental factors in determining "sex" for many species means that 'sex' is a much more flexible concept (Snowdon, 1997).

Although discussions of chromosomes tend to focus on the genes inherited from the father (because it is these genes that define the sex of the child) all people must have at least one X chromosome because so many essential genes are contained on the X chromosome. All fetuses spend their first six weeks in an XX womb and her amniotic fluid, undergoing the same development until the release of testosterone for the majority of XY fetuses.[22] The only thing that does not exist is a pure (Y or YY) male. There has been a case of a boy born with an XX configuration, however. This boy's ovum split several times before being fertilized by sperm, providing further evidence that parthenogenic reproduction extends to humans. All cells usually contain a conglomeration of our biological parents' chromosomes. This means that our bodies live in a permanently fertilized state, with only our egg and sperm cells qualifying as sexed (haploid): the vast majority of our cells are intersex (diploid).

The determination of what we tend to understand as "sex" (i.e. the appearance of genitals and the presence of ovaries or testes) is not determined by the X and Y chromosomes alone. What is known in biology as the "testis-determining factor" (TDF) works on the fetus's developing gonad. If present, the gonad will develop into testes; if absent the gonad will develop into ovaries. In mice, for instance, if the gonads of an XY mouse were placed in an XX mouse, the mouse would develop into a male mouse, despite having XX chromosomes (LeVay, 1993). Indeed, the relationship between genes and hormones is complex. LeVay outlines the conditions necessary for particular genes to be activated by the gonads: "(1) that gene must possess the characteristic DNA sequence allowing it to be bound by a particular steroid receptor, (2) the gene must not have been inactivated by some other overriding process (such as occurs, e.g. when large blocks of genes, useless for a particular tissue or cell type, are permanently switched off during development), (3) the particular steroid receptor must be present in that cell, and (4) the steroid itself must be present in sufficient concentration, which means either that it must be present in sufficient levels in the bloodstream, or that the cell must contain converting enzymes capable of creating it from some other steroid that *is* present in the blood" (1993: 25–6).

Investigations into the workings of genes are far from over. Genomic studies have thus far failed to deliver on the promise to uncover the "secret of life." Yet, these same studies provide much more interesting data about the interdependence of DNA with its environment. As for predictions of the future of genomic studies, Fox Keller makes the following:

> Only three predictions seem safe to make about the character of biology in a post-genomic age. First, a radically transformed intra- and intercellular bestiary will require accommodation in the new order of things, and it will include numerous elements defying classification in the traditional categories of animate and inanimate. (2000: 9)

Conclusions

To sum up, the major analytical thread which connects all of these diverse studies is that discussions of "matter" are socially mediated. Feminist studies of skeletons, gonads, hormones, and genes emphasize that these so called signifiers of sex "differences" are far from being either primordial or immutable. Feminists emphasize that these physical, material aspects of the body do not express meaning in themselves, but are rather made meaningful within a social discourse that structures differences between women and men. In this way, each study raises an important issue about the relationship between the cultural and the physical. As Simon LeVay points out, "biology is *inseparable* from its contextual meaning ... nature *cannot be* separated from nurture' in the way that traditional arguments about "nature *versus* nurture" suppose. As I argued in Chapter 1, this does *not* mean that feminist theory should abandon science studies. It does mean that we want to take a fresh look at scientific research. As we will see in the next chapter, more recent science studies speak of the interaction between nature and the environment, rather than attempting to impress upon any analysis the exclusivity of nature or the environment.

Suggested readings

Keller, E.F. (2000) *The Century of the Gene*. Cambridge, MA: Harvard University Press.

Martin, E. (1991) "The Egg and the Sperm: How Science Has Constructed a Romance Based on Stereotypical Male-Female Roles," *Signs: Journal of Women in Culture and Society*, 16(3): 485–501.

Oudshoorn, N. (1994) *Beyond the Natural Body. And Archaeology of Sex Hormones*. London and New York: Routledge.

Peterson, A. (1998) "Sexing the Body: Representations of Sex Differences in Gray's *Anatomy*, 1858 to the Present," *Body and Society*, 4(1): 1–15.

4
New Materialism, Nonlinear Biology, and the Superabundance of Diversity

Man [sic] has been here 32,000 years. That it took a hundred million years to prepare the world for him is proof that that is what it was done for. I suppose it is. I dunno. If the Eiffel Tower were now representing the world's age, the skin of paint on the pinnacle knob at its summit would represent man's [sic] share of that age; and anybody would perceive that the skin was what the tower was built for. I reckon they would, I dunno.

(Mark Twain in Gould, 2000: 45)

This means – and we must face the implication squarely – that the origin of *Homo sapiens*, as a tiny twig on the improbably branch of a contingent limb on a fortunate tree, lies well below the boundary [between predictability under invariant law and the multifarious possibilities of historical contingency]. In Darwin's scheme, we are a detail, not a purpose or embodiment of the whole ...

(Gould, 2000: 291)

It seems that natural selection shows no concerns for our labels of what is masculine and what is feminine.

(Breedlove in Snowdon, 1997: 280)

Introduction

There are a series of dualities in academia so persistent as to have the appearance of philosophical immortality: agency versus structure in sociology; nature versus nurture in psychology; good versus evil in religion and philosophy; and the public versus private in political theory.

In Chapter 2, I borrowed Matt Ridley's observation that "similarity is the shadow of difference: difference is the shadow of similarity" (2003: 7), to make the point that it is equally plausible to discuss "sex" similarity as it is to discuss "sex" difference; that we might argue for the relative strength of one factor or the other, but that both factors constitute sides of the same proverbial coin. In this chapter, I want to examine another duality that persists, explicitly or implicitly, throughout the literature on evolutionary theory: conformity versus diversity. Here, I refer to conformity as a conservative quality in the extent to which the morphology and behavior of living organisms are confined by law-like parameters dictated by nature. In contrast, diversity refers to the extent to which morphology and behavior express a wide range of characteristics produced through the principle of variation.

Evolutionary theory is commonly assumed to favor sexual reproduction over nonsexual reproduction and sex differences over sex diversity. These assumptions, however, are based more on competing evolutionary theories than on Darwin's original thesis. For instance, functionalist evolutionary theories lead to a conception of "sex" as having a particular function and that, in turn, this function is, or produces, sex complementarity. New materialism, on the other hand, has generated a renewed interest in what I argue have become more silent, yet nevertheless intrinsic, elements of Darwinian theory: contingency, diversity, nonlinearity, and self-organization (all of which are distinctly nonfunctional). In this chapter I review those aspects of Darwin's theory that particularly relate to "sex" diversity. Against an exclusive emphasis on the immutability of sex "differences," I argue that evolution equally evinces diversity, contingency, and variation. I am not arguing that evolutionary theorists do not acknowledge these important elements of evolution, but rather that public understandings of sex "differences" are anchored by a skewed understanding of the principles of evolutionary theory. If "difference is the shadow of similarity," then diversity is the shadow of conformity, and new materialism is a promising mechanism through which this shadow may be rendered more vivid. The chapter concludes by exploring how some feminist theorists are using the principles of new materialism to look at materiality as both active and positive. I explore these feminist analyses of materiality as a prelude to the remaining four chapters of the book that examine various aspects of the materiality of sex diversity. As such, the present chapter signposts the basic tenets of both evolutionary theory and new materialism; Chapters 5, 6, 7, and 8 explore how these theories might be applied to questions of "sex" and sex "differences."

Evolutionary theory

Today, evolution has entered into common vocabulary and interpretations of its founding text, both scholarly and populist, vary tremendously. To read the *The Origin of Species* (1859/1998) is therefore to grasp an opportunity to estimate how closely that common understanding relates to its source. There are often vast differences in what Darwin wrote, and how his work has been used in what has become known as Social Darwinism or Sociobiology.

Like all radical scholars, Charles Darwin's work needs to be situated within its cultural context. At the time that Darwin was developing his theories about the workings of living organisms in nature, Christianity dominated European social thought. Both ancient and Christian social thought conceived of the universe as structured by a set of static, immutable hierarchical relationships between all beings (God, angels, the Sovereign, human beings, animals, and the earth), each with a distinct purpose. Human behavior was understood to be determined by an overall, God given, natural and moral structure and purpose, and premodern social treatises were mainly concerned with outlining the elements of an ideal society based on this moral structure.

Beginning in the fifteenth century, major advances in science through the works of Copernicus, Kepler, Galileo, and Newton began to introduce a radical challenge to the dominant Christian view. In stark contrast to the Christian belief in the hierarchical, purposeful, and static relationship between living and nonliving beings, the revolution in science "conceived of the universe as a mechanical system composed of matter in motion that obeyed natural laws. Both divine purpose and human will became peripheral, indeed unnecessary, features of the scientific world view" (Seidman, 1994).

In *The Origin of Species* Darwin drew upon evidence derived from horticulture, domestic breeding, and fossil records. Darwin noted that detailed records kept of the domestic breeding of animals and plants revealed that many breeds did not always exist; that, indeed, the purpose of domestic breeding was to produce, over successive generations, animals and plants better suited to human needs. The production of useful breeds was predicated on a simple principle – to successively *select* variations from each generation of animal or plant that most resembled the desired breed. Darwin summarized the process thus: "this preservation of favorable variations and the rejection of injurious variations, I call *Natural Selection*" (1998: 64 emphasis mine). In short, Darwin argued that human breeding of animals and plants mimics what nature does over millions of years.

As Stephen Jay Gould argues, the "bare-bones mechanics of natural selection is a disarmingly simple argument, based on three undeniable facts: *overproduction of offspring, variation,* and *heritability"* (2003: 13, see also Waage and Gowaty, 1997). That is, Darwin argued that species typically produce more offspring than are necessary for survival (which he termed "superfecundity"), that natural variations occur with regularity in organisms, and that organism characteristics are inherited. Gould further outlines three principles underlying natural selection. *Agency* refers to the principle that single organisms are the locus of selection, such that any appearance of the "good design" of organisms or the harmonization of ecosystems is an unintended consequence.[23] *Efficacy* refers to the principle that variation slowly, over a tremendous number of generations, leaves the fit and eliminates the unfit. Finally, *scope* refers to the principle that the entire diversity of species on the planet is explained by microevolutionary processes over an immense timescale. In order to capture the process of evolution, both in terms of the natural selection of characters and the momentous timescale, Darwin and many subsequent evolutionary theorists pictorialized evolution as a branching tree. The "root" or bottom of the tree represents the beginning of life, from which (after millions of years) "branches" of the tree represent the diversification of life into different classifications such as insects and mammals.[24]

Natural selection has a number of corollaries. First, natural selection cannot in itself produce favorable variations, which must occur by chance. Moreover, evolution also occurs through genetic drift recombination and mutation (see Chapter 5). Second, what proves to be a favorable variation is dependent upon the environment. This means that a variation favorable in one environment might well be unfavorable in another environment (both geographic and temporal). This principle has important implications for the survival of species, which may thrive in one set of environmental conditions but flounder and ultimately perish (become extinct) in another set of conditions, a theme I will return to in the next section. The point here is that we cannot know which species (if any) will ultimately prevail. In other words, natural selection does not have the power of foresight, and so cannot create adaptations to prevent extinction (Waage and Gowaty, 1997).

For example, population growth and viability are very strongly affected by environmental variability. Environmental variability includes such factors as predation, temperature, and resource availability, and may affect populations in terms of number and health of offspring, reproductive rate, as well as infant, child, and adult mortality. Researchers are therefore keenly interested in the degree to which individuals and populations demonstrate plasticity in variable environments (Komers, 1997). Organisms display plasticity to the extent to which they can change their phenotypes in accordance with environmental changes.

Second, variations that are neither useful nor harmful would not be affected by natural selection and would occur randomly throughout any species over time (Gould, 2000). We should expect variations in physical traits to be correlated with variations in levels of survival fitness, except in cases where a particular character is favorable in present environmental conditions but evolved for other reasons, or where the character evolved in tandem with another character that increased species fitness (Sork, 1997: 109). Third, natural selection works on an extremely long timescale through the accumulation of infinitesimally small inherited modifications, each profitable to the organism (Gould, 2000).

In sum, Darwin's theory of natural selection radically overturned the doctrine of a divinely ordered structure of beings in the world.[25] Nature, over millions of years, created the diversity of living organisms. Moreover, this diversity was generated from an original set of living forms; species evolved rather than remaining in their original form since their creation. Further, natural selection is an extremely long process; new organic beings are not simply "created"; nor are there any sudden or momentous modifications to characters. Finally, perhaps the most difficult pill to swallow was the idea that humanity is not the "end-goal" or "purpose" of evolution. In the "tree of life" representation of evolution, human beings clearly evolved very recently in geological time – the myriad of life which flourished before the evolution of human beings did not exist "because of" human beings, an important point I will return to later in the chapter.

Before leaving this short synopsis of evolutionary theory, it is worth drawing attention to two common confusions about the mechanism of natural selection. Jean-Baptiste Lamarck (2004) theorized that behaviors useful to organisms could be transmitted to offspring during their lifetime; a theory later proved to be false by the hereditary condition of evolutionary theory. The second confusion relates to the erroneous notion that characters evolve because they are functional to the organism. The sheer volume of anomalies in the living world contests this idea. For instance, Gould (1987) uses the example of human male nipples. Male nipples may provide pleasure from stimulation for some men, but in strictly survival terms, they do not serve any function, and Gould argues this is precisely the point – "male nipples may have not adaptive explanation at all" (1987: 16). As Chapter 5 of this book will show, females and males are not separate "opposing" beings, but rather slight variants, shaped together by natural selection, of "a single ground plan" (1987: 16). The point made earlier is that variations that are not maladaptive will not necessarily disappear through the natural selection process.

Sociobiology and sexual selection

Since Darwin devised his radical theory in *The Origin of Species*, hundreds of evolutionary scientists have expounded on natural selection and evolution. Edward Wilson coined the term "sociobiology" to describe a branch of evolutionary theory concerned with "the systematic study of the biological basis of all social behavior" (Wilson, 2000: 4). Drawing data mainly from nonhuman animal groups, but also early human animal cultures, sociobiology focuses on aspects of social behavior such as altruism, communication, dominance, roles and castes, social spacing including territories, parental care, animal homosexuality (see Chapter 6), and behavior seen to be related to sex such as parental care, monogamy, and sex selection. As Chapter 1 indicated, sociobiology has been subjected to major criticism from social scientists, and particularly feminists, because it often appears to provide a biological explanation for culturally conservative practices that reinforce unequal "sex" roles in society. For instance, much sociobiological emphasis is placed on competition and dominance in studies of animals, with inferences from the interpretation of observations of animals made to human society. Wolfgang Wieser (1997) observes that studies focus on competition rather than cooperation because this reflects the Western emphasis on aggression and competition in human social systems. These studies clearly associate dominance with the males of most species and passivity with females. However, what we see in our culture is not necessarily the result of "nature" through evolution. In other words, we must be cautious in justifying social conditions through recourse to nature. As Sork argues:

In order for a character to continue to evolve by natural selection, that character must have a fitness advantage and a heritable basis. Due to the evolution of human culture, many characters that provide a fitness advantage may no longer have any heritability. Thus, we would not expect a response to natural selection in the next generation. In fact, many of the kinds of characters that are often discussed as the basis for unequal status (e.g., leadership, political authority, parental skills) are the ones that are most vulnerable to difficulties in measurement and to the influence of environmental conditions. (1997: 106)

Sociobiology focuses on two topics that are particularly germane to this book: parental investment and the evolution of "sex" itself; topics critiqued more extensively in Chapters 5 and 6. The point to be made here is that sociobiology tends toward conformity as outlined at the outset of this chapter. Recall that conformity refers to a conservative quality in the extent to which the morphology and behavior of living organisms are confined by law-like parameters dictated by nature. Sociobiology tends to emphasize the corroboration of human cultural expectations of sex complementary behavior such as maternal care and protection of offspring, male competition for females, male interest in numerous sexual partners, male aggression and female passivity, and so on. For this reason, new materialism offers an alternative perspective on the "nature" of "sex" and sex "differences."

New materialism and nonlinear biology

A number of social scientists dissatisfied with the propensity of socio-biology to reinforce a conformist view that the morphology and behavior of living organisms are confined by law-like parameters dictated by nature have begun to turn to new materialism and nonlinear biology more specifically, to explore diversity. A number of social theorists note a significant shift in the natural sciences away from an emphasis on determinism to a recognition of "open-endedness and of emergence" (Grosz, 1999c: 19). New materialism has for some time moved toward an understanding of matter as a complex open system subject to emergent properties. In *One Thousand Years of Nonlinear History* (2000) Manuel De Landa traces the history of the philosophy of matter to demonstrate how simple behavior, defined through the emerging science of chemistry as matter that conforms to the laws of definite properties, became the major focus of scientific attention. Tremendous gains were made in understanding properties of inert matter, but slowed the recognition of what Gilles Deleuze and Felix Guattari (1987) term the "machinic phylum," or the overall set of self-organizing processes within the universe, including organic and inorganic matter, that is produced by nonlinear dynamics (De Landa, 1991). De Landa explains Deleuze and Guattari's concept of the machinic phylum in terms of how different structures (geologic, biological, socioeconomic) are produced through "strata" (homogeneous elements such as sedimentary rocks, species, and social hierarchies) and "meshworks" (heterogeneous elements such as igneous rocks, ecosystems, and precapitalist markets) (De Landa, 1997a: 509).[26] New materialism emphasizes three interrelated, critical concepts: nonlinearity and self-organization, contingency, and variation or diversity.

Nonlinearity and self-organization

The nonlinearity and self-organizing properties of matter are predicated on an understanding of the agency of matter itself. The following is one of the best descriptions I have ever read of the agency of matter. Writing in 1883, Mark Twain describes the Mississippi river:

> The Mississippi is remarkable in still another way – its disposition to make prodigious jumps by cutting through narrow necks of land, and thus straightening and shortening itself. More than once it has shortened itself thirty miles at a single jump! These cutoffs have had curious effects: they have thrown several river towns out into the rural districts, and built up sand bars and forests in front of them. The town of Delta used to be three miles below Vicksburg: a recent cutoff has radically changed the position, and Delta is now *two miles above* Vicksburg. Both of these river towns have been retired to the country by that cutoff. A cutoff plays havoc with boundary lines and jurisdictions: for instance, a man living in the State of Mississippi today, a cutoff occurs at night, and tomorrow the man finds himself and his land over on the other side of the river, within the boundaries and subject to the laws of the State of Louisiana! Such a thing, happening in the upper river in the old times, could have transferred a slave from Missouri to Illinois and made a free man of him. (1961: 14–15)

In this example, the physical dynamics of the Mississippi produced changes in the social dynamics of race relations, and new materialism emphasizes the emergence of historical and evolutionary change through networked interaction of human and nonhuman entities. But while matter self-organizes, it is not toward any other goal than itself. *Self-organization* refers to the absence of any underlying program controlling development. Stent uses the example of ecological succession:

> Bare rock is exposed as a consequence of volcanic or glacial action. After the rock has weathered sufficiently to allow the formation of some soil lichens to colonize the surface and accelerate the formation of soil, these interactions create a suitable environment for grasses and herbs to colonize, and they eventually replace the lichens. Larger

shrubs and trees follow the colonization of this environment by grasses and herbs providing a suitable habitat for herbivorous animals. Once herbivores are present, then omnivorous and carnivorous organisms can then colonize the developing community. (Gray, 1997: 396)

Science presents an increasingly detailed picture of matter as a self-organizing, networking, complex system of emergent organic and non-organic properties. Kevin Kelly outlines these emergent properties to include maximizing heterogeneity which "speeds adaptation, increases resilience, and is almost always the source of innovations" (1994: 604). Related to heterogeneity is the principle of seeking persistent disequilibrium as the "continuous state of surfing forever on the edge between never stopping but never falling" (1994: 605). Also included is the principle of honoring errors: evolution itself is "systematic error management" (1994: 605). Finally, Kelly argues that emergent properties pursue no optima, but have multiple goals:

An adaptive system must trade off between exploiting a known path of success (optimizing a current strategy), or diverting resources to exploring new paths (thereby wasting energy trying less efficient methods). So vast are the mingled drives in any complex entity that it is impossible to unravel the actual causes of its survival. Survival is a many-pointed goal. Most living organisms are so many-pointed they are blunt variations that happen to work, rather than precise renditions of proteins, genes, and organs. (1994: 605)

Put another way, matter is nonlinear. In linear systems, there are only two kinds of possible long-term dynamics: either the system stays constant, or it grows or declines at a constant rate. In contrast, at least three types of nonlinear dynamics are found in biology. According to Ferrière and Fox (1995), dynamical systems on limit cycles repeat themselves regularly. Quasicycles resemble limit cycles, except that the periods of oscillations vary, such that the system never identically repeats itself. Finally, chaotic oscillations do not exhibit regular periods or amplitudes, making it impossible to predict the system's long-term behavior (and hence the name "chaos"). An example of a nonlinear system is population size because populations vary without either growing infinitely or disappearing. De Landa notes that nonlinear dynamics have been found at the biochemical, organism, and population levels, and that, more broadly, nonlinear changes must affect the evolution of character traits. Indeed, most biological models are nonlinear, which again suggests evolutionary change through networked interaction of human and nonhuman entities.[27]

Emergent hybridizations are not solely the product of human agency, but are indigenous to networking open systems. De Landa characterizes this nonlinear history as "a narrative of contingencies, not necessities, of missed opportunities to follow different routes of development, not of a unilinear succession of ways to convert energy, matter, and information into cultural products" (1997b: 99). In short, evolution has no foresight – it is not headed toward perfection. Evolution is better characterized, in the words of Arthur Koestler, as "epic tale told by a stutterer" (in De Landa, 1997b: 71). As De Landa argues, " 'natural selection' ... is merely the failure of all possible offspring that are born or hatched to persist and reproduce in the game of life" (1997b: 71).[28] Social scientists are interested in nonlinearity not merely as the absence of telos but insofar as it emphasizes the "accidental, chance, or the undetermined plays in the unfolding of time" (Grosz, 1999c: 18).

Contingency

Perhaps the most radical theme of new materialism is the principle of contingency. Recall that the beginning of the chapter recounted that evolutionary theory overturned the Christian paradigm of organic and inorganic matter organized in purposeful harmony in the ultimate service of humanity. New materialism, following evolutionary theory, argues the opposite. Coining the term "organic chauvinism" De Landa emphasizes that if nature has a "point," it is the process itself, not the coagulation of nature (of which our bodies are a prime example):

> In the eyes of many human beings, life appears to be a unique and special phenomenon ... This view betrays an 'organic chauvinism' that leads us to underestimate the vitality of the processes of self-organization in other spheres of reality ... In many respects the circulation is what matters, not the particular forms that it causes to emerge ... Our organic bodies are ... nothing but temporary coagulations in these flows: we capture in our bodies a certain portion of the flow at birth, then release it again when we die and micro-organisms transform us into a new batch of raw materials. (De Landa, 1997b: 103–4)

Although the branching "tree" of evolution discussed earlier in the chapter illustrates the central features of evolution as both the natural selection of characters on a momentous timescale, this tree figure has been criticized insofar as it suggests that human beings constitute the apex of evolution (note that Darwin himself did not depict this hierarchy). For instance, the branching of the tree from a common root in the ground suggests that the branches at the upper end of the tree represent the most advanced species, or that each branching limb represents an equal level of diversity of life forms within a particular taxonomy of living organisms. However, as Gould argues, evolution is not a story of slow steady progress toward the perfection of species, but rather a story of "decimation by lottery" (2000: 261). He further points out that:

> Evolution is not a sequence of progressive replacements rooted in superior anatomy on an eternal battleground. Reptiles did not replace fishes; rather, they represent an oddly modified group of fishes in a novel terrestrial environment. Fishes have never been replaced as dominant vertebrates of the oceans. (2000: 259–60)

In *Wonderful Life* (2000), Gould reanalyzes the fossil evidence of the famous Burgess Shale on the west coast of Canada to argue that an enormous number of species were decimated (by environmental conditions) followed by the vast differentiation of the few surviving species. Like De Landa, Gould spells out the implications for understanding humanity's place in evolution:

> We cannot bear the central implication of this brave new world. If humanity arose just yesterday as a small twig on one branch of a flourishing tree, then life may not, in any genuine sense, exist for us or because of us. Perhaps we are only an afterthought, a kind of cosmic accident, just one bauble on the Christmas tree of evolution. (2000: 44)

Both De Landa and Gould point to one of the central tenets of evolution; that natural selection does not mean progress (something that Darwin himself struggled with because he understood the moral implications of his theory within a Christian paradigm. Darwin wrote "After long reflection, I cannot avoid the conviction that no innate tendency to progressive development exists" (in Gould, 2000: 257)). Natural selection responds to changes in local environments, rather than to an overall plan or design. When the environment changes, genetic mutations do not "respond" to the environment because "variation itself supplies no directional component" (Gould, 2000: 228).

For this reason, the notion of contingency is so pivotal within new materialism. Gould defines contingency as "the tendency of complex systems with substantial stochastic components, and intricate nonlinear interactions among components, to be unpredictable in principle from full knowledge of antecedent conditions, but fully explainable after time's actual unfoldings" (2002: 46). Contingency refers to the idea that evolution is not the march of species' progress through time, but rather the result of an indefinable number of contingencies. Gould provides a humorous example:

> Even if fishes hone their adaptations to peaks of aquatic perfection, they will all die if the ponds dry up. But grubby old Buster the Lungfish, former laughing-stock of the piscine priesthood, may pull through – and not because a bunion on his great grandfather's fin warned his ancestors about an impending comet. Buster and his kin may prevail because of a feature evolved long ago for a different use has fortuitously permitted survival during a sudden and unpredictable change in rules. And if we are Buster's legacy, and the result of a thousand other similarly happy accidents, how can we possibly view our mentality as inevitable, or even probable? (2000: 47)

Buster survives because of the contingency of life. Gould titled his own book on contingency after the Frank Capra film *It's a Wonderful Life*. In the film, a despondent George Bailey (played by Jimmy Stewart) is visited by an angel who shows George what the town of Bedford Falls would have been like without his presence. In this replaying of the life of Bedford Falls without George, the angel encapsulates the principle of contingency when he exclaims, "Strange isn't it? Each man's life touches so many other lives, and when he isn't around he leaves an awful hole, doesn't he? ... You see, George, you really had a wonderful life." Just as the lives of the citizens of Bedford Falls are contingent on George Bailey's life, so too are species contingent on a highly complex interaction of environmental factors. Different environmental conditions produce different evolutionary results. Contingency is one of the central tenets of nonlinear biology, as Evelyn Fox Keller notes:

> Biology is not lawful in the same sense in which physics is, for every feature of a biological organism is what it is by virtue of its long evolutionary history. And the reason the outcome of all these 'billion years of experimentation by its ancestors' is never either absolute or predictable is that the experimental materials with which primitive life forms could work were themselves dependent on the occurrence of chance events. Life as we know it is the beneficiary of this long

history of fortuitous opportunities. Stephen Jay Gould likens evolution to a videotape that, if replayed over and over, would have a different ending with each playing. In fact, it is sometimes argued that chance, or contingency, is the defining characteristic of evolution, and possibly even its driving force. (2000: 103)

Gould's metaphor of the videotape, that when played over and over again will produce a different ending every time, is at the crux of the contingency principle that new materialism places so much emphasis on. As Gould argues:

> The issue of prediction, a central ingredient in the stereotype, does not enter into the historical narrative (neither does verification by repetition since we are trying to account for uniqueness of detail that cannot occur together again). We can explain an event after it occurs, but contingency precludes its repetition, even from an identical starting point. (2000: 278)

Variation and diversity

Finally, new materialism emphasizes the principle of variation or diversity. As Snowdon writes, variation:

> is the raw material of natural selection. Without the variation produced by mutation, recombination, genetic drift, and behavioral plasticity, there would be no change, no need to write or think about evolutionary biology. Diversity, individual variation, and change are of greater importance than stasis or consistency. (1997: 276)

Random genetic drift occurs when random events in the lives of individuals influence the "fitness" of that individual (fitness here being defined solely as the production of offspring).[29] As Collins notes "apparently well-adapted individuals can get unlucky and leave few offspring in the next generation, and apparently poorly adapted individuals can get lucky and leave many offspring" – recall Buster the Lungfish (1994: 246). Random processes, chance in other words, influence genetic changes in populations without natural selection. One corollary is that sexual reproduction is not necessary for variation (see Chapter 5).[30] For example, mutations may become either fixed or eliminated in populations through the process of genetic drift. The same may be said for behavioral expressions. Although biologists are typically trained to look for adaptive (i.e. naturally selective) reasons why particular behaviors flourish and others perish in a given population, Thelma Rowell (1979) argues that traditions and cultures may well drift in the same random fashion. Recall how natural selection works in the first place. Natural selection requires diversity in order for it to act. Natural selection is a long, continuous process of selecting out various forms which have an advantage in the present environment (and these same forms may not have an advantage when the environment changes). In her review of literature on the social organization of primates, Rowell argues that it is extremely difficult to "distinguish between differences which are truly adaptive and those produced by random drift ... and it therefore behooves us not to accept all differences as having selective advantage without bearing the alternative [random drift] in mind" (1979: 14).[31]

Thus, new materialism aims not to distill matter's incalculable variation to a simple, single explanation of "reality," but rather to normalize these very differences (Ferguson, 1997: 10). Indeed, anthropologist Paul Rabinow argues that if "nature" is to "retain any meaning at all it must signify an uninhibited polyphenomenality of display" (Rabinow, 1992: 249).

Feminists intra-acting with matter

Taking into account the emphasis of new materialism on emergent properties, contingency, variation, and diversity has helped a number of feminist theorists to think about materiality without the usual accompaniment of essentialism, where matter is understood as an inert container for outside forms. The remainder of this chapter provides a flavor of the kinds of analyses feminist scholars are undertaking using new materialism.

One of the reasons I think feminists are increasingly engaging with Gilles Delueze and Felix Guattari's work is because nature is not conceived under a "juridical transcendent plane" (i.e. in need of translation and governance by humans) but as immanently self-organizing (Gatens, 2000: 60). Deleuze and Guattari (1983, 1987) have developed a theory that, in refiguring matter as molecular, mobile and dynamic, challenges theories that figure bodies as solid inert objects as well as distinctions between human and nonhuman, and, living and nonliving matter.

Deleuze and Guattari's theory of "becoming molecular" stands in rather stark contrast to feminist theories of the body as excluded "other"

(Irigaray, 1985; Moi, 1986). Because representations of the body and "sexual difference" are seen to be effects of a prerepresentation (usually the maternal feminine), analyses tend to be negative – a mourning for that which is lost through masculinist representation. Deleuze and Guattari, on the other hand, do not consider bodies as vehicles of consciousness or as privileged sites of meaning (Bray and Colebrook, 1998: 56). As Abigail Bray and Claire Colebrook write:

> Matter, or the body, would not be thought's 'other' if thinking were seen as a desiring production, a comportment, an activity, or an ethos. The body is not essentially anterior or other. And it follows from this that a theory of sexual difference that relies on constitutive negation may be best overcome by not turning to the body or attacking representation but by questioning the primacy of the representation/ materiality dichotomy. (1998: 56)

Deleuze and Guattari present an account of materiality such that the body is a positive event rather than a negated origin; "action is productive rather than representational" of some originary lack (Bray and Colebrook, 1998: 57. See Colebrook, 2000a for a provocative delineation of the metaphysical differences between Irigaray and Deleuze).

Thinking about bodies as "becomings" rather than the signification of an originary loss has encouraged a number of feminists to consider "sexual difference" differently. Some feminist scholars have taken up Deleuze and Guattari's work to question notions of "sexual difference" as prerepresentation, such as Elizabeth Grosz's edited collection *Becomings* (1999a), Ian Buchanan and Claire Colebrook's *Deleuze and Feminist Theory* (2000), and the special issue of *Hypatia*, "Going Australian: Reconfiguring Feminism and Philosophy" (2000). For instance, in *Becomings*, Grosz has collected a number of works that emphasize concepts of chance, randomness, and open-endedness. These articles attempt a distinctly nonsocial constructionist account of becoming, insofar as social constructionism is dependent upon human interpretation to open up the world. Grosz describes "upheavals going on in biology and biological modeling [as] more akin to the randomness of evolution, the unfolding of lineage and mutation" (1999c: 28). She links the concept of becomings with biology insofar as "biology has opened itself to futurity and thus aligned itself with certain contemporary physicists' notion of indeterminancy" (1999: 20). Buchanan and Colebrook's collection is also concerned with the notion of becoming, and employs a Deleuzian analysis to explicitly contest the dependence of traditional social constructionist arguments on human interpretation. In this collection, Colebrook asks "what if sexual difference thought itself as a problem ...? If philosophy were neither a question of the opening of pure truth, nor a question of recognition, but the confrontation of new problems and concepts, then sexual difference would be a different form of difference" (2000a: 124). Colebrook employs a Deleuzian philosophy of the event, rather than the concept (such as the originary subject found in some cultural analyses) to argue that difference does not generate the subject, and thus "sexual difference is no longer foundational, no longer the difference from which all other (given) differences are effected" (2000a: 118).

A number of feminists have also analyzed "sexual difference" as origin through the work of Jacques Derrida. Feminist theory's use of Derrida's writing is wide-ranging. But to my mind, one of the most interesting and thought-provoking ways in which Derrida's insights have been harnessed is in thinking about new materialism. A hallmark of Derrida's work is his concern with the notion of origin or pure presence. For Derrida, the world is constituted through signs, inscriptions, and marks that are subject to repetition. In order for each sign to be discernible from the next, it must *differ* in some way, that is "an interval, a distance, *spacing* must be produced between elements" (Derrida in Grosz, 1986: 34). This spacing or interval is also a movement in time, such that each iteration of the sign is *deferred*. For Derrida, then, each iteration effects both continuity and difference, or *différance*, so as to obviate any notion of an origin or absolute self-identity.

Yet Kirby observes that while cultural deconstruction may claim to have dismantled the notion of origin, the very division between "nature" and "culture" reinstates an origin:

> But what is happily relinquished in the critique of the subject is then quietly recuperated elsewhere. The identity of the subject as an atomic principle of indivisible autonomy has certainly been sacrificed, appearing in qualified form as an 'emergence' within a generalized field of becoming. The explanatory force that can no longer be ceded to the subject, or indeed to any identity, has nevertheless miraculously resurfaced in the entity of 'culture' itself. (1999: 21)

Kirby explores how the work of Derrida has been selectively taken up by feminists to further cultural analyses at the expense, according to Kirby, of potential Derridean applications to materiality. She asks "why is Derrida's privileging of 'writing' and 'language' read as cultural constructivism par excellence, as if Nature is placed under erasure by

Culture" (1999: 20)? As Annemarie Jonson writes, Derrida himself is critical of this reduction of writing to culture:

> Indeed, while Derrida remarks not infrequently on the 'differential and formal character of semiological functioning', the 'possibility of code ... independent of any substance', he elsewhere stresses equally the deconstructive insistence of materiality. For matter, in Derrida's view, is philosophy's debased 'exterior', a 'radical alterity ... in relation to philosophical oppositions'; and insofar as material substance may be thought 'outside the oppositions in which it has been caught (matter/ideality, matter/form) ...', he concedes that 'what [he] write[s] may be considered 'materialist'. (1999: 56)

Drawing out some observations initiated by Kate Soper (1995), Kirby challenges the traditional nature/culture opposition as an example of Derridean supplementarity, whereby those sociological theories that attempt to ground "nature" entirely within "culture" grant to "nature" an extra-discursive order of reality. Kirby argues that a Derridean analysis effects a "mediated nature of nature [that] neither nature nor culture can accommodate comfortably" (1999: 25). Kirby uses DNA and cells as material examples of languages that should prompt a rethinking of language within the nature/culture divide. In this sense, the "text" that Derrida claims has no outside is the "text of nature" (1999: 28). Here is a thesis that suggests the restriction of some poststructural and postmodern deconstruction to discourse has implications for both nature and science, and Kirby argues that we must open up the "text" to outside determinations not concerned with the human subject (2001).[32] Derrida's notion of *différance* invites an understanding of nature and culture as concepts which are neither pure presence nor absolute self-identity, but indicate instead marks within a "web, a textile of other marks, a mesh of constantly moving parts extending through time and space" (Clark, 2001: 96). Vicki Kirby's analysis of Derridean "text" as DNA takes feminist analysis along a path seldom traversed by most feminist theory – which remains firmly anchored to an implicit separation between culture and matter.

Rosalyn Diprose (1991) provides a further excellent illustration of feminist engagement with new materialism. Challenging the widely held assumption that the genetic code is the origin of biological (and sexual) differences, Diprose goes beyond cultural analyses of genetics that argue against a causative relation between genetic codes (genotype) and their expression (phenotype). Developing Derrida's theory of the origin as always deferred, Diprose argues that the genetic code has no *material* origin. Specifically, genetic codes are determined by the pairing and ordering of nucleotide bases, that is, their relation to each other. Moreover, DNA codes only become operative when they replicate into a mirror image of themselves and then reverse this process, not back to an originary code but to "the other of the other" (1991: 72). As Elizabeth Wilson explains:

> The process of DNA–RNA transcription effects a double deferral: from the nucleotide bases to their interval, and from their interval to a series of transcriptions that never return to the origin. It is the processes of spacing, difference, and translation without original – rather than the repetition of the same from a present origin – that determines genetic effect. It is this trace of a trace, rather than a present and locatable code, that is the genetic 'origin'. (1998: 99–100)

Diprose demonstrates matter at play to show that even genetic difference is not grounded anywhere but rather produces its own origin as effect.

Also drawing upon the insights of new materialism, Karen Barad (1998, 2001) offers an analysis that combines critical theory with physics, to propose an epistemology for comprehending "things" (matter) that does not depend on a notion of "truth as a faithful reflection of a static world of being" (De Landa, 2000:2). Barad develops what she terms "agential realism" to refer to (among other things) the nature of scientific and other social practices, the nature of reality, the nature of matter, and the relationship between the material and the discursive in epistemic practices. Agential realism seeks to move beyond the traditional division between "realism" and "social constructivism." Whereas classical Newtonian physics assumes that observations can be transparent (that a distinction can be made between observations and objects), Niels Bohr argued this distinction to be impossible. Bohr defined a "phenomenon" as the lack of inherent distinction between objects and their agencies of observation (Barad, 2001: 231). This means that "reality is not composed of things-in-themselves or things-behind-phenomena, but things-in-phenomena" (Barad, 2001: 235). This ontology does not suppose being as prior to signification (as in classical realism and some cultural feminist theory), but neither does it understand being as a product of language (as in some cultural formulations). Rather, agential realism examines the ways in which nature and culture *intra-act* as, for example, how different disciplinary cultures (such as feminist theory) define what counts as "nature" and what counts as "culture" (2001: 240).

A number of feminist scholars concerned with science studies, and nonlinear biology specifically, offer interesting and useful ways of *intra-acting* with matter. For instance, Sarah Franklin argues that the most pervasive and powerful representation of nature is as a biological entity; that the origin of "life itself" is represented in biological terms as natural selection, and egg and sperm activity (2000). Franklin traverses conflicting representations of, on the one hand, biology as *telos* of organic survival through sexual reproduction – sociobiological accounts such as Richard Dawkins's *Selfish Gene* (1989) – to, on the other hand, the nonlinearity of genes as information reproduction. One of the significant implications of the shift to "genomic governmentality" is that "many of [biology's] former foundational fictions are now in the reliquary beside Lamarckism, [and] neither life nor sex [are branches] on the same family tree that Darwin borrowed from the Bible to begin with" (Franklin, 2000: 219).

Like Barad, Donna Haraway develops a notion of materiality as both material and semiotic effect. Haraway is particularly interested in transspecies/cendence/fusions/gene/genics/national that disturb the hierarchy of taxonomic categories (genus, family, class, order, kingdom) derived from pure, self-contained and self-containing "nature." For Haraway, trans "cross a culturally salient line between nature and artifice, and they greatly increase the density of all kinds of other traffic on the bridge between what counts as nature and culture" (1997: 56). What appeals to me about the concept of "trans" is that it works equally well both between and within matter, confounding the notion of the well-defined, inviolable self which precedes Western culture's "stories of the human place in nature, that is, genesis and its endless repetitions" (1997: 60). As Haraway argues, in these Western stories "history is erased, for other organisms as well as for humans, in the doctrine of types and intrinsic purposes, and a kind of timeless stasis in nature is piously narrated. The ancient cobbled-together, mixed-up history of living beings, whose long tradition of genetic exchange will be the envy of industry for a long time to come, gets short shift" (1997: 61).

Haraway (2001) goes on to provide a superb example of how knowledge of biological diversity can inform key feminist debates about embodiment and "the self." Haraway describes *Mixotricha paradoxa*, a minute single-celled organism that lives in the gut of the South Australian termite. This tiny organism engenders key questions about the autonomy of identity (we tend to assume that single organisms are defined by the possession of nucleated cells), or as Haraway puts it "the one and many" (2001: 82). *Mixotricha paradoxa* lives in a necessary symbiotic relationship with five other organisms, none with cell nuclei but all with DNA. Some live in the folds of the cell membrane, while others live inside the cell, while simultaneously not being completely part of the cell. Haraway asks: "is it one entity or is it six? But six isn't right either because there are about a million of the five non-nucleated entities for every one nucleated cell. There are multiple copies. So when does one decide to become two? And what counts as *Mixotricha*? Is it just the nucleated cell or is it the whole assemblage?" (2001: 82). Advancing a similar argument, Joost Van Loon (2000) uses symbiosis theory within nonlinear biology to argue the parasite with the body as the ultimate "Other," and invites a reconsideration of a politics of difference from inside the body (see also Rackham, 2000).

A number of feminists are forefronting analyses of the kinds of intra-action that Barad and Haraway refer to. The 1999 issue of *Australian Feminist Studies* features a series of articles on feminist science studies. Guest editor Elizabeth Wilson sets out the current state of feminist analyses of science with the following quote: "that culture, history and language exist at all must, in some broad sense, be in the nature of human biology" (1999: 7). Wilson argues that a feminist audience is likely to attribute to this quote a determinist or essentialist author. The fact that the author is Elizabeth Grosz might surprise many feminist scholars of "the body," and Wilson is interested in exploring how feminist analyses have proceeded "as though the nature of biology is immaterial" (1999: 7). Using hysteria as an example of the "psycho/somatic," Wilson argues that feminist analyses of corporeality have narrowed to such an extent as to be completely reduced to cultural discussions. Wilson calls for feminists to contemplate biology, bodies, and matter as the important, but mainly overlooked or erased, details of corporeality, and suggests that feminist theory will remain impoverished insofar as it does not attend to these details.[33]

Elizabeth Grosz's article *Darwin and Feminism: Preliminary Investigations for a Possible Alliance* (1999b) encourages feminists to use the analytic tools provided by evolutionary theory – abundant individual variation, proliferation of life forms, and the "play" of natural selection – to analyze such diverse themes as oppression, social change, relations of sexual and racial difference, as well as a number of dualisms that have frustrated feminists, including the apparent nature–culture divide. As Grosz argues, "evolution is a fundamentally open-ended system which pushes toward a future with no real direction, no promise of any particular result, no guarantee of progress or improvement, but with every indication of inherent proliferation and transformation" (1999b: 39). This means that culture is not the end product of nature, or any sort of logical culmination or going-beyond of nature. In this sense, nature cannot be overcome by culture, or culture by nature, as they are one in the same process. A number of articles in this special issue go on to consider specific 'matterings' such as Helen Keane's analysis of the addicted brain, Catherine Waldby's analysis of The Visible Human Project and ideas about reproducing life, Adrian Mackenzie's analysis of technology, Celia Roberts's review of Haraway's "material-semiotic" actors, and Anna Munster's analysis of cyberfeminism. Annemarie Jonson evaluates the dualism between "form" and "matter" in artificial life studies that hold "form" (in this case computer program) to be the "essence" of life. But rather than provide a more traditional analysis of the social construction of this dualism, as, for instance, feminist analyses interested in the ways in which form and matter became gendered, Jonson turns to molecular biology. Like Diprose (1991) Jonson shows how the interaction among genes (as "computer program" form), proteins, and cytoplasm is not a unidirectional set of instructions that cause proteins to function in certain ways, but rather a complex interaction dependent upon the environment.

Conclusions

Jonathan Waage and Patricia Gowaty (1997) argue that the "nature–nurture" debate is misguided because it ignores the fundamental principle of evolution. They write "successful current designs (adaptations) reflect the relative past performances of previous and coexisting designs that were organic responses of individuals in populations to problems imposed by their past environments" (1997: 586). This means that, in evolutionary terms, environment and genotype are coextensive with one another and not mutually exclusive as the traditional debate would have it. This chapter has sketched the outlines of Darwinian evolutionary theory – natural selection over millions of years – and argued that new materialism may be seen as an extension of evolutionary theory insofar as it emphasizes what seems almost lost in the hegemony of sociobiology in public discourse: self-organization and nonlinearity, contingency, variation, and diversity.

Confining feminist criticism to cultural practices naturalizes the distinction between cultural and material domains, reduces the complexity of biological matter that new materialist studies emphasize, and reauthorizes the very practice of segregating "feminist concerns" from "neutral" explorations that anchor traditional scientific endeavors. I see in these "feral publications" (to use Vicki Kirby's term) an unbridled enthusiasm absent in more mainstream feminist cultural analyses, which seem at times to be founded upon a resolute determination to find all that is negative in science and matter. It is almost as if the polyphony of living and nonliving matter has infected these feminist scholars, who are proving to be an "enthusiastic audience ... hungry for more curious and excited modes of feminist interaction with the sciences" (Wilson, 2000: 40. See also Grosz, 1999b, c).

In this brief account, I have tried to indicate the flavor of these theories, which delve into matter to understand feminist concerns with culture, and use new materialism to consider questions such as the origin of sexual difference. But these explorations go further: by bringing DNA and nonhuman matter within the purview of feminist critique, the project of feminist theory is at the very least expanded to include areas traditionally considered outside of feminist consideration because they do not explicitly refer to women or "sexual difference". At most, these critiques are a "breach ... against conventionalization ... the infraction of immobile boundaries and a displacement of the fixed political–critical spaces they enact" (Wilson, 1998: 204). The remaining chapters of this book devote themselves to contributing to the feminist critique of "sex" and sex "differences" through the analytic lens of new materialism.

Suggested readings

De Landa, M. (1997b) *A Thousand Years of Nonlinear History*. New York: Swerve Editions.
Gould, S.J. (2000) *Wonderful Life. The Burgess Shale and the Nature of History*. London: Vintage.

5
The Nonlinear Evolution of Human Sex

> The very fact that nonsexual reproduction is called asexual reveals the normative preference given to sexual reproduction.
>
> (Schiebinger, 1993: 22)
>
> Not that it really matters whether or not he [sic] ever knows about the vast populations of inorganic life, the 'thousand tiny sexes' which are coursing through his veins with a promiscuity of which he cannot conceive. He's the one who misses out. Fails to adapt. Can't see the point of his sexuality. Those who believe in their own organic integrity are all too human for the future [to come].
>
> (Plant, 1997: 205)

Introduction

Chapter 3 examined skeletons, gonads, hormones, and genes as often-cited signifiers of sex "differences" between women and men. However, the ability of some women to sexually reproduce is the most frequent and powerful signifier of "sexual difference" in Western cultures. Whatever social, political, and economic changes might take place to alter women's position in society, female sexual reproduction is seen as both immutable "fact" and cause of structural differences between women and men. Of the almost countless references to female "materiality" as reproduction, my training as a sociologist secures Emile Durkheim's rendition as a particularly sharp thorn in my side. He writes, "... society is less necessary to her because she is less impregnated with sociability ... Man is actively involved in it whilst woman does little more than look on from a distance" (1970: 385). Not only does Durkheim remind his readers that it is female bodies that can be (passively) impregnated, but this impregnation is limited to fleshy materiality (babies). If male bodies are (actively) impregnated, it is with decidedly nonmaterial sociality.

All feminist texts contest the oppression experienced by women through cultural assumptions about sex "differences," but few texts actually contest the basis of the claim that sex "differences" exist. The aim of this chapter is to challenge the *a priori* acceptance of "sexual difference" based upon sexual reproduction. I purposefully use the specific term *sexual* reproduction here instead of the more commonly used generic term reproduction. As we will see in this chapter, all living bodies reproduce: *autopoesis*, or the self-maintenance of an organism, is one of the fundamental organizing principles of life. Take the human liver, for instance – our bodies reproduce liver cells about every two months. *Sexual* reproduction is a much more specific enterprise: it requires that the reproduction taking place involves sexual intercourse or *relations* of some sort between living organisms of *different* sexes.

The acceptance of the idea that sexual reproduction is the basis of "sexual difference" is based on three widely held assumptions. The first assumption is that sexual reproduction is the most common form of reproduction among living matter. Second, it is assumed that sexual reproduction has an evolutionary purpose. Finally, we tend to assume that the human body itself is sexually differentiated. Drawing upon data from new materialism and nonlinear biology, this chapter will challenge each of these assumptions. I will argue that the current recourse to "the body" based upon reproductive function selectively attends to one aspect of "materiality" – that is, human bodies (like all other living organisms) engage in constant and varied reproduction, and only a small proportion is sexual.

Girding the association between sexual reproduction and "sexual difference" is the assumption that sexual reproduction is the basis of associations of kinship. The chapter concludes by introducing chimerism and mosaicism, two forms of reproduction that take place in human populations (and in nonhuman populations) that radically challenge cultural notions of the genetic or "blood" basis of kinship. This discussion will draw on analyses of xenotransplantation and reproductive technologies to argue that suppositions about the "natural" basis of kinship are reliant on cultural imperatives rather than founded in "the nature of things." That is, reproductive technologies, chimerism, and mosaicism challenge cultural assumptions insofar as sexual reproduction is assumed to be the basis of kinship.

Reproducing bodies

The most compelling representation of a nonlinear system in which multiple forms of matter–energy (including minerals, biomass, and genes) enter into nonlinear relationships with uncertain outcomes, is the body (Clark, 2000). In this section, I want to offer some resources for thinking about reproduction in a nonlinear biological frame.

Interacting bodies

As the previous chapter outlined, traditional evolutionary theory constructed a system of hierarchical relationships between, and within, plant and animal species. However, contemporary nonlinear biology understands this relationship as much more of a meshwork than a top–down or bottom–up system. And replacing the traditional two-kingdom classification, scientists now speak of five: bacteria, protests, fungi, plants, and animals.[34] Most of the organisms in four out of the five kingdoms do not require sex for reproduction (Margulis and Sagan, 1997). Species within these kingdoms interact in dynamic ways, each with the potential to change each other's adaptive environment. For example, only a very few primitive fungi are two sexed. Schizophyllum, on the other hand, has more than 28 000 sexes. And sex among these promiscuous mushrooms is literally a "tough-and-go" event, leading Laidman to conclude that for fungi there are "so many genders, so little time ..." (2000: 1–2).

Only by taking our skin as a definitive impenetrable boundary are we able to see our bodies as discrete selves. Our human bodies, like those of other animals and plants, are more accurately "built from a mass of interacting selves. A body's capacities are literally the result of what it incorporates; the self is not only corporeal but corporate" (Sagan, 1992: 370). The cells in our bodies engage in constant, energetic reproduction. Oyama refers to this "mobile exchange" of genetic, intra and extracellular and environmental influences as a "choreography of ontogeny" (Jonson, 1999: 51). Indeed, the millions of microbes which exist on, and in, our bodies makes our traditional definition of ourselves as single organisms highly problematic, an important point to be extended later in this chapter. Our cells also provide asylum for a variety of viruses and countless genetic fragments. And none of this interaction requires any bodily contact with another human being.

More than 50 synthetic chemicals flow into our bodies daily (including tinned vegetables, cigarettes, chemical detergents, makeup, DDT) and alter our endocrine systems (Colborn, Dumanoski, and Myers, 1996: 199). Cosmic irradiation, the acquisition of viruses and symbionts, and exposure to chemicals also alter, or add to, our DNA structure, which produces variation without sexual reproduction (Margulis and Sagan, 1997). Endocrine-disrupting compounds have been found to be responsible for a recently reported doubling in incidence of hypospadias in the United States and Europe (Dolk, 1998; Paulozzi *et al.*, 1997). Children are at risk of exposure to over 15 000 high-production-volume synthetic chemicals; most of them developed in the last 50 years. More than half have not yet been tested for toxicity (Landrigan *et al.*, 1998). The effects of DDT and DDE have been studied on a diverse range of animals from Tiger Salamanders to Cricket Frogs (Clark, Norris, and Jones, 1998; Reeder *et al.*, 1998). A number of researchers are interested in the possible causal relationship between exposure *in utero* to environmental chemicals and effects on human sexual reproduction including sex ratio, disruption of androgen signaling, decreased sperm number and quality, androgen insensitivity, testicular and breast cancer, decreased prostate weight, endometriosis, decreased fertility, increased hypospadias and undescended testes, as well as adverse effects on immune and thyroid function (Cheek and McLachlan, 1998; Golden *et al.*, 1998; Olsen *et al.*, 1998; Santti *et al.*, 1998; Skakkeæk *et al.*, 1998; Tyler *et al.*, 1998).[35] Again, each of these exchanges with the environment may effect variations in sex and fertility without any recourse to sexual reproduction.

Evolving bodies

Not only have evolutionary biologists replaced the two-kingdom schema with five-kingdoms, but the major division is no longer between plants and animals, but between eukaryotes (cells with nuclei such as plastids and mitochondria) and prokaryotes cells lacking membrane-bounded nuclei, such as bacteria (Margulis and Sagan, 1997).[36] Human beings evolved from the protist lineage. And protists developed mitotic sex; one of the most common forms of reproduction whereby cell division takes place by maintaining the chromosome number. Thus, during most of our evolutionary heritage, our ancestors reproduced without sex.

Not only, as I have said, do we tend to think that reproduction on this planet requires sex, but a pervasive heteronormative (see Chapter 2) assumption claims that "sex" *must* have some evolutionary purpose. But as Margulis and Sagan (1991) argue, "sex" may have no evolutionary purpose whatsoever. The mere existence of any particular anatomical trait (the appendix is the most commonly cited example) does not mean this trait was an adaptation in the interests of survival. Indeed, many evolved traits are either neutral or maladaptive. Margulis and Sagan (1986) argue that sexual reproduction evolved by accident as a necessary

by-product of the evolution of multicellularity and cellular differentiation. They argue that it is not sexual reproduction that makes humans
so successful, but rather multicellularity. By observing a host of different
cell types, Margulis notes that when cells differentiate, they lose their
ability to divide. In multicellular organisms, cells begin to specialize and
carry out different functions. Some cells in multicellular organisms
specialize in sexual reproduction and become known as the germ line.
These germ cells both divide and provide the means by which to pass
on the cellular genomes in the cytoplasm and the nucleus to the next
generation. In order to differentiate, the organism must retain some
cells that use *meiosis* (halving the number of chromosomes – referred to
as a *haploid* state) to ensure that accurate copies of its genes are passed
on to the next generation. Meiosis requires fertilization in order to return
to a *diploid* (doubling of chromosomes to make a complete set) state.
Margulis and Sagan write that "mixis ... becomes a consequence of the
need to preserve differentiation ... mixis itself is dispensable and ... was
never selected for directly" (1986: 180). Put another way, "multicellularity provided evolutionary advantages and sex came along for the ride
(Fausto-Sterling, 1997: 53)".

Margulis and Sagan's theory emphasizes the concept of randomness,
chance, and contingency, factors that play a significant role in nonlinear
evolution (see Chapter 4). Thus, rather than deliberate on how most
living organisms are able to reproduce without "sex," scientists are more
puzzled by those species which *do* engage in sexual reproduction. Sexual
reproduction consumes twice the energy and genes of parthenogenic or
asexual reproduction (Bagemihl, 1999: 254). After an extensive search
on the biological literature on sex, Mackay concluded:

> The most intriguing aspect of my research was why we have sex at
> all. After all, sexual reproduction in animals started only 300 million
> years ago. Life on earth got on pretty well for 3000 million years
> before that with asexual reproduction ... [Sexual reproduction] takes
> more time, it uses more energy, and mates may be scarce or uncoop-
> erative. (2001: 623)

In "The Cost of Mating," Martin Daly outlines why sexual reproduction
"presents a paradox for the theory of natural selection" (1978: 771).
Daly notes that meiosis (or sexual reproduction) as opposed to *parthenogenesis* (or nonsexual reproduction) only ensures that about 50 percent
of any offspring's genes will be inherited (more for females). This paradox has vexed evolutionary biologists, who have spent many years

analyzing the cost–benefit of sexual reproduction, and largely conclude that sexual reproduction is actually "maladaptive" if it were placed in competition with parthenogenic reproduction. The benefit of sexual reproduction is referred to as *mixis*, which means that any offspring that have been sexually reproduced inherit a combination of genetic information. A great number of scientists have postulated hypotheses as to why this combination of genetic information might be evolutionarily superior to parthenogenic reproduction. The *parasite hypothesis*, for instance, suggests that faster evolution is produced through sexual reproduction, which helps host species who are in constant competition with parasite species (Collins, 1994). In a constant system, parasites have the advantage over host species because parasites reproduce more quickly, thus allowing faster evolution.

But there are many costs to sexual reproduction. The *twofold genetic cost* refers to the fact that sexually reproducing individuals only contribute about half of their genetic information to each offspring. A parthenogenic female, on the other hand, contributes all of her genetic material to her female offspring, who will in turn contribute all of their (and their mother's) genetic material to their offspring, and so on. This parthenogenic female would double the frequency of her genotype in each generation, and her lineage would require few generations to take over the population (Collins, 1994: 252). Another advantage of parthenogenic reproduction (and thus another cost of sexual reproduction) is what is known as the *twofold ecological cost* of sex. Nonproductive males contribute little or nothing (besides their gametes) to their offspring and represent a drain on the ecosystem – nonproductive males (i.e. males who do not contribute resources to their offspring) use up half of the species' ecological resources. A parthenogenic female only reproduces females, and so she does not waste resources on males.

The costs of sexual reproduction also include the scarcity of mates, the orchestration of the activities of two organisms; the competing interests of these two organisms, a greater amount of time required for "courtship," copulation, and gestation; the possible competition of rival mates; as well as the energetic costs of the mechanics of having sexual intercourse, sexual behavior, and escape from unwanted sexual attention. Add to this list the risks of predation, disease transmission, and injury inflicted during sex on both the female, the male, and any dependent offspring (see Hrdy, 1974). Interestingly, Daly makes the point that in species lacking paternal investment, there is particularly little benefit for the female to engage in sexual reproduction. A female in such a species that could reproduce parthenogenically could "dispense with all of the costs [listed] above, as well as those of meiosis and recombination, while losing nothing in male aid, since there is none to lose" (1978: 773). Daly concludes that sexual reproduction is merely an evolutionary vestige which will be supplanted should parthenogenesis evolve in species that currently use sexual reproduction.

Reproducing bodies

In contrast to the minimal amount of specifically sexual reproduction that some human beings engage in, each of us engages in constant reproduction. Thus apart from the fusing of separate bodies, human beings engage in *recombination* (cutting and patching of DNA strands), *merging* (fertilization of cells), *meiosis* (cell division by halving chromosome number, for instance in making sperm and eggs), and *mitosis* (cell division with maintenance of cell number). Margulis and Sagan refer to "jumping genes, 'redundant' DNA, nucleotide repair systems, and many other dynamic genetic processes [that] exploit the 'cut and paste' recombination of ancient bacteria-style sexuality that evolved long before plants, animals, or even fungi or protists appeared on this planet" (1997: 181). Moreover, we constantly reproduce our own bodies as an essential feature of autopoiesis. Not only do we reproduce our own livers every two months, but also our stomach linings every five days, new skin every six weeks, and 98 percent of our atoms every year (Margulis and Sagan, 1995: 17).

Our human bodies live in a permanently fertilized state, with only our egg and sperm cells qualifying as sexed (haploid): the vast majority of our cells are intersex (diploid). As Chapter 3 outlined, 44 of our 46 chromosomes are completely unrelated to sexual difference. The only thing that does not exist is a pure (Y or YY) male. Recall that there has been a case of a boy born with an XX configuration, however.

Donna Haraway highlights the key irony of our evolving and reproducing bodies, that in biological terms sex precludes reproduction:

> There is never any reproduction of the individual in sexually reproducing species. Short of cloning ... neither parent is continued in the child, who is a randomly reassembled genetic package projected into the next generation. To reproduce does not defeat death any more than killing or other memorable deeds of words. Maternity might be more certain than paternity, but neither secures the self into the future. In short, where there is sex, literal reproduction is a contradiction in terms ... Sexual difference founded on compulsory heterosexuality is itself the key technology for the production and perpetuation of western Man and the assurance of this project as a fantastic lie. (1989: 352)

This irony, that any child is a "randomly reassembled genetic package," contrasts sharply with cultural notions of kinship. Indeed, whether it is the pre-twentieth century notion of "blood ties" or contemporary emphasis on genetic inheritance, kinship remains perhaps the central concern in discussions of maternity, paternity, and sexual reproduction more generally. As such, the remainder of this chapter explores challenges to cultural notions of kinship from a new materialist perspective.

Sexual reproduction and kinship

Perhaps the strongest structure of exclusion (and, by definition, inclusion) is kinship. Donna Haraway's definition of kinship, as "a technology for producing the material and semiotic effect of natural relationship, of shared kind" (1997: 53) reinforces the dependence of notions of kinship on culture rather than nature, and highlights the refraction of biological assumptions about "blood kinship" through cultural notions of "blood kinship." Much recent work in feminist studies of science, and the sociology of science more generally, has expanded analyses of the ways in which culture influences biological notions of kinship, for instance in studies of reproductive technologies (Donchin, 1989; Franklin, 1997, 2001), the Visible Human Project (Waldby, 1999), animal studies (Haraway, 1989, 1997), and intersex (Fausto-Sterling, 2000; Hird, 2000, 2003), all of which are explored elsewhere in this book. In this section, I want to focus on some perhaps less well-known biological phenomena that raise interesting questions about sexual reproduction and kinship.

Chimerism refers to the presence of two genetically distinct cell lines (genomes) in an organism. This may occur through inheritance, transplantation, or transfusion. Mosaicism is more common than chimerism and refers to patches of tissue that differ genetically.[37] Chimerism is most familiarly known within both the xenotransplantation literature, as the transplantation of animal organs into humans, and the non-human animal literature, where chimerism has been documented in a large range of nonhuman animals including cats, mink, dogs, horses, pigs, cattle, sheep, goats, primates, rabbits, rodents, and chickens. Chimerism has entered scientific discussion more recently with respect to genetic inheritance in humans (Bird *et al.*, 1982; Nelson, 2002; Neng *et al.*, 2002; Pearson, 2002; Strain *et al.*, 1995, 1998; Van dijk, Boomsma, and de Man, 1996). Once assumed to be a rarity, recent research (e.g. Van dijk, Boomsma, and de Man, 1996) suggests that as many as 4 percent of human twins and 14 percent of triplets are chimeras, as well as a yet unknown incidence within the general population (one does not need to be born as a multiple birth – chimerism may occur because of an embryo "that died early in gestation and was spontaneously aborted") (Pearson, 2002: 10).

What I aim to argue here is that chimerism both, as xenotransplantation, and the more recent "discovery" of chimerism as blood and genetic transfer, incite interesting analyses of kinship. Chimerism introduces a curious paradox – on one hand chimerism as xenotransplantation *broadens* the notion of kinship to include relationships between human and nonhuman animals. On the other hand, chimerism and mosaicism within the new biological literature offer the opposite – to *contract* traditional understandings of "blood" and genetic relations such that a mother may not be "blood" related to the children that she gives birth to, and individuals may share germ cell lines with never-living siblings. In this analysis I want to argue a point that Marilyn Strathern (1992) has made, that nature does not provide a sufficient model for the cultural context of kinship. I will suggest chimerism and mosaicism may serve to remind us that our cultural conceptions of what kinship means, and what biology "says" are not transparent or immutable. Cultural notions of sexual reproduction are closely related to kinship; we assume one produces the other. But this is not necessarily the case at all. Recall Haraway's earlier insight that the project of kinship is, biologically speaking, founded on a "fantastic lie" because sexual reproduction ensures that the self is never faithfully reproduced in offspring due to the fact that offspring consists of a randomly reassembled genetic mixture. Chimerism and mosaicism provide yet more examples of the ways in which sexual reproduction does not lead to kinship in the way we culturally imagine.

All in the family

Kinship is most often defined within the context of traditional Western heteronormative society. Within this structure, kin is dichotomous – either blood or non-blood relations. Blood relations are assumed to share biological substance including genes and blood, and to have resulted from sexual reproduction. The study of kinship has been one of the great mainstays of anthropology, and during the heyday of the anthropological tradition of extensive ethnographic study of non-Western cultures, anthropologists discovered that some cultures used classificatory "blood" kinship terms that did not correspond to what were thought by Euro-North American anthropologists to be "true" genetic relationships (i.e. biological). Trobriand Islanders and Aboriginals of Australia, for instance, deployed a complex system of relations to define "kin," some based on what Euro-North American anthropologists recognized as "blood relations," and some based on "non-blood relations." In effect, Trobriand Island and Aboriginal systems of kinship challenged Euro-North American assumptions about the consanguinity of kinship. As Franklin observes, "it was a perception that derived from the European scientific assumption that kinship categories should be read directly from 'blood' ties as a matter of commonsense, and that to do otherwise could only be interpreted as ignorance of paternity, or general lack of intellectual development" (1997: 22). In his famous studies of the Trobriand Islanders, for instance, Bronislaw Malinowski argued, "it seems hardly necessary to emphasize that for physiological consanguinity as such, pure and simple, there is no room in sociological science" (1913: n. 177).

In contrast, David Schneider sought to analyze the ways in which American culture has become increasingly dependent upon notions of biology. In his path-breaking work on kinship in *American Kinship: A Cultural Account* (1968), Schneider offered a sustained account of the complex relationship between biology and kinship. Just 12 years later the hegemony of biology had become such that, in the second edition of the book, Schneider was able to argue that:

In American cultural conception, kinship is defined as biogenetic. This definition says that kinship is whatever the biogenetic relationship is. If science discovers new facts about biogenetic relationship, then that is what kinship is and was all along. (1980: 23)

Insofar as Schneider argues that biology has no meaning outside of cultural context, he highlights the particular contradictions of Euro-North American understandings of kinship. Schneider argues:

The relationship between man [sic] and nature in American culture is an active one ... Man's place is to dominate nature, to control it, to use nature's powers for his own ends ... In American culture man's fate is seen as one which follows the injunction Master Nature! ... But at home things are different. Where kinship and the family are concerned, American culture appears to turn things topsy-turvy ... What is out there in Nature, say the definitions of American culture, is what kinship is. ... To be otherwise is unnatural, artificial, contrary to nature. (1968: 107)

Schneider's work has been the subject of critical analyses that have pointed out, for instance, that it relies upon a distinction between "cultural facts" and "biological facts" at the same time that it seeks to expose this distinction in other anthropological work (Franklin, 1997). Nevertheless, Schneider's focus on heterosexual coitus as the central "symbolic universe" of American kinship has been taken up within contemporary lesbian, gay, and reproductive technology kinship studies. Franklin notes that "amid the many transformations that have reshaped the study of kinship over time, the question of the significance of biological facts has remained a persistent quagmire – as easy to fall into as it is difficult to leave behind" (2001: 302).

I want to focus now on analyses of reproductive technologies insofar as these technologies challenge public imaginations of kinship, especially in cases of intergenerational gestation, genetically related egg and/or sperm donation, and nongenetically related egg and/or sperm donation. I will argue that in unintended ways, chimerism and mosaicism work in tandem with reproductive technologies to challenge Euro-North American understandings of kinship based on sexual reproduction.

Reproductive technologies offer a number of challenges to the traditionally constructed linear equation of sexual reproduction: that heterosexual coitus leads to pregnancy which leads to offspring of direct kin relation to her/his parents. In the first instance, as Franklin observes, for the "growing number of couples ... [for whom] coitus *never* results in pregnancy, or for whom even conception and implantation do not result in pregnancy, the usefulness of the 'biological model' is ... in question" (1997: 64). For subfertile or infertile heterosexual individuals, coitus very rarely results in pregnancy. Moreover, in the majority of cases where reproductive technologies are used, conception and implantation of embryos also does not result in pregnancy. So right from the start, traditional understandings of kinship fall far short of the reality for many heterosexual people, as well as lesbian and gay people. Add to this the growing use of sperm and egg donation, and the traditional understanding of kinship is further challenged.

In the case of reproductive technologies, a woman who uses egg donation might gestate and give birth to a child she has no genetic relationship with (or more specifically, no genetic relationship through the egg – genetic material does transfer through blood). Or a woman who uses the egg of her own mother might give birth to a child who is, genetically speaking, her sister. The list of variations goes on (see Thompson, 2001). In each of these cases, we might argue that kinship is *extended* beyond traditional "criteria" to include more than the person who gives birth to a child and her partner (i.e. to include egg donor, sperm donor, and so on).

Carlos Novas and Nikolas Rose observe that "new reproductive technologies have split apart categories that were previously coterminous – birth mother, psychological mother, familial father, sperm donor, egg donor and so forth – thus transforming the relations of kinship that used to play such a fundamental role in the rhetoric and practices of identity formation" (2000: 490–1). As Charis Thompson observes, "biological motherhood is becoming something that can be partial" (2001: 175). That is, reproductive technologies invite such emotive concern from the public because these technologies demonstrate that biogenetics underdetermines kinship, insofar as kinship is defined as both primordial and immutable. In this way, just as anthropologists found that "primitive" cultures use classificatory systems, we could well argue that Western cultures use these same classificatory systems, even while they depend upon strong notions of "biology." That is, we *assume* that mother and child are blood related, that children do not share germ cells with their never born siblings. But these common assumptions may not be corroborated by biological evidence.

Recent research investigating chimerism and mosaicism further confound traditional notions of embodiment. Studies in biology refer to chimerism as the presence of two genetically distinct cell lines in an organism. This may occur through inheritance, transplantation, or transfusion. For instance, recall the boy who was recently born in Britain who is, genetically speaking, two people because he was formed by the fertilization of two eggs and two sperm which then fused into one embryo (Pearson, 2002). It is becoming increasingly clear that cells traffic between fetus and mother in both directions during pregnancy, and those fetal cells continue to circulate for years in the mother after birth. This "microchimerism" has also been found in multiply transfused recipients of blood transfusions (Nelson, 2002). Other interesting examples have turned up in the medical literature. The cell and tissue blood of one boy had none of his father's chromosomes, but did have a duplicated set of one half of his mother's chromosomes (Pearson, 2002). In another case, a mother was discovered not to be the genetic mother of her four children (whom she had gestated and given birth to, and had not used donor eggs). This woman has two populations of genetically different cells, one in her blood and the other in her gonads, and that only the cells in her gonads were transferred to her children. As noted above, mosaicism is more common than chimerism and refers to patches of tissue that differ genetically. This would result in a person having two genetically distinct cell lines on a part or parts of their body. Like reproductive technologies, chimerism and mosaicism introduce challenging variations to traditional notions of kinship. In some cases chimerism and mosaicism produce a similar extension of kinship criteria – for instance, to never-living siblings. But sometimes they produce the opposite – these biological variations *contract* kinship such that a woman who uses her own egg, uterus, and blood to produce a child might not be "blood" or genetically related to this child. A man whose sperm is used to fertilize an egg that produces a child may not be "blood" or genetically related to this child.

What is so interesting about chimerism and mosaicism is that whereas public understandings of reproductive technologies are deeply imbued with concerns about tampering with "nature," chimerism and mosaicism are "natural" in the sense that they have undergone no human technological intervention (except in cases of transfusion or transplantation). Chimerism and mosaicism may be viewed as "anomalies" but they stand outside of human technological intervention even as they fundamentally challenge traditional notions of kinship. As Franklin observes, "ideas of the natural comprise one of the most important 'cultural logics' that more recent theorists of kinship and gender have sought to analyze" (1997: 57). And in analyzing this "cultural logic" we find that "nature" and "science" are deployed in uncomfortably contradictory ways. What chimerism and mosaicism demonstrate is that *nature* can contradict the *cultural* assumption that children are biologically related to their (nonadoptive) parents, at the same time that this cultural assumption is supposed to be grounded in biological explanation. It is for this reason that Franklin and McKinnon (2001) argue that the privileging of kinship rests on a tautology. Moreover, while the interpreter of what is "natural," science, is imbued with characteristics of rationality and impartiality within Western traditions, science may also reveal relationships where none are assumed (between living and never living siblings), and no relationship (between mother and child) where such a relationship is the foundation of kinship systems.

Boundaries – inclusion and exclusion

Science and technology enjoy an ambivalent position in the cultural imagination engendered, as the previous chapter argued, by anxieties about the coherence and stability of human being. And as the above discussion aimed to elucidate, processes of inclusion and exclusion are at the heart of cultural configurations of kinship. These processes entail the establishment of boundaries which serve to exclude certain notions of embodiment threatening to the human sense of self. For instance, Kath Weston asks:

> If kinship can ideologically entail shared substance, can transfers of bodily substance create – or threaten to create – kinship? Can they create – or threaten to create – other forms of social responsibility? What investment do people have in depicting the transfer of blood, organs, and sperm as sharing, giving or donation? What investment do they have in resisting such transfers (or the vehicles of transfer)? Alternatively, how do people work to construe transfers as 'signifying nothing' with respect to race, sexual contact, religious identity, and so on? (2001: 153)

Just as reproductive technologies threaten established understandings of kinship (of inclusion and exclusion) Kath Weston is arguing that science and technology offer both the promise *and* threat of new configurations of selfhood, responsibility, and kinship.

Prior to the research I have outlined in this chapter on chimerism and mosaicism in human blood, skin cells and genes, chimerism was already courting center stage within research on xenotransplantation. At the launch of the joint report on xenotransplantation grafting by the British Union for the Abolition of Vivisection and Compassion in World Farming, one scientist warned:

> The human xenotransplantation patient will become a literal *chimera* ... It sounds like scare-mongering, but let me assure you that the word *chimera* is being used by xenotransplantation scientists ....
> (Quoted in Brown 1999a: 191, my emphasis)

Xenotransplantation involves the use of nonhuman animal cells or organs in human animals. We may think of the concept of kinship not only in terms of intraspecies inclusions and exclusions (as in the case of human animals) but also between species. Thus Weston's questions about the boundaries of kinship do not just apply to the transfers of human organ and tissue between humans, but extend to these transfers between human and nonhuman animals.

A great deal of boundary work is done to continually distinguish between human and nonhuman animals.[38] Xenotransplantation engenders public concern to the extent that it threatens to collapse notions of interspecies kinship boundaries. Donna Haraway explores this boundary collapse in her work on biogenetic relationships such as OncoMouse™ that create a specific form of genetic relationship between humans and mice (humans and mice are, of course, already genetically related. So are humans and bananas for that matter). These "trans" relationships "simultaneously fit into well-established taxonomic and evolutionary discourses [for instance, technological progress] and also blast widely understood senses of the natural limit" (Haraway, 1997: 56). Franklin notes that "the ways in which humans are today connected and related through biology *undoes the very fixity that the biological tie used to represent*" (2001: 314 original emphasis).

Debates invoked by xenotransplantation are heavily dependent upon an implicit notion of the monster, in this case in terms of the "authentic" boundaries of the self (especially here in terms of humans versus nonhumans). Nik Brown states:

> The monster variably expresses: the repressed forces buried beneath the promethean purpose ... ; critiques of utopian rationality and the overwhelming pace of industrial capitalism ... ; the insecurities of human ontology represented in the disorder that rises from Victor's disgust at an artifice which traverses death and life, organism and machine ... (1999b: 341)

The "pollution" created by this "monstrous" transgression of boundaries requires action: "the delineation of a border, the naming of transgressors, the ritual of the purge, the subsequent restoration of a boundary" (Brown, 1999b: 342).[39]

Thus public concerns about chimerism as xenotransplantation and reproductive technologies can be understood as contemporary distillations of kinship boundary work. Xenotransplantation and reproductive technologies effectively extend traditional understandings of kinship as "flesh and blood." And both technologies do so through an explicit and primary use of notions of "nature" and science; the very same notions the Western concept of kinship has relied upon to define (through exclusion) itself.

As well as highlighting cultural ambivalences about the boundaries of kinship, and the limits of human selfhood and being, public concerns about xenotransplanation and reproductive technologies reveal a selective use of biological evidence. For instance, public concerns about the "pollution" of xenotransplantation leaves out what should be a parallel discussion of the "pollution" already in human bodies, that is present even before birth. As noted in Chapter 4, Joost Van Loon (2000) uses symbiosis theory to argue the parasite within the body as the ultimate "Other," and invites a reconsideration of a politics of difference from inside the body. Human bodies, like those of other animals, live in necessary transspecies symbiotic relationship.[40] As Alphonso Lingus recognizes:

> ... human animals live in symbiosis with thousands of anaerobic bacteria – six hundred species in our mouths, which neutralize the toxins all plants produce to ward off their enemies, four hundred species in our intestines, without which we could not digest and absorb the food we ingest ... The number of microbes that colonize our bodies exceed the number of cells in our bodies by up to a hundredfold. (1994: 167)

Concomitantly, Jami Weinstein argues, "given the biological reality that without the constant dynamic interaction between human bodies and the autonomous bodies of other living organisms human bodies would not survive, the economy of the independent, unitary, fixed, stable, whole body becomes a fantasy (or a fiction of science)" (2003: 308).

If particular relationships, such as those engendered by reproductive technologies and xenotransplantation, were to be placed in a hierarchy according to the degree to which they challenge traditional understandings of kinship, then the acknowledgment of transspecies interdependence represents perhaps the apex of such a hierarchy. Moreover, however radical this conceptualization might seem to public discourses of kinship, it does not depend upon any arguments about "new" technologies. The necessary symbiotic relationships that human animals engage in are not the result of what are usually thought of as technologies that humans have created, and so bypass arguments that are made with regard to xenotransplantation and reproductive technologies, which are erroneously seen as entirely human developed. Franklin notes that there has been an overestimation of both the novelty and determinism of human technological innovation – the "plus ça change argument ... may well serve as an important counterweight to the overreaction that may occasion developments" such as reproductive technologies and xenotransplantation (2001: 319).

And this recognition can only serve to reinforce the sociological and anthropological argument that Western notions of kinship, while explicitly reliant upon "natural facts," are implicitly imbued with contradictory, unsettled, anxious, and ever fracturing cultural discourses. Chimerism and mosaicism are yet further examples of biology destabilizing cultural understandings of biology (in this case, the "monster" turns out to be the very knowledge base, nature, which public discourses ironically use to exemplify "the natural"). As Michael and Carter argue, "... science tends to emerge narratively as a 'parasite' that disrupts the smooth circulation of practices and discourses that comprise local social identities" (2001: 12).

Conclusions

This chapter has argued that sexual reproduction is not a reliable signifier of "sex" or "sex differences" for the following reasons. First, not all women sexually reproduce: up to 30 percent of the world's female population does not sexually reproduce. Second, reproduction does not have much to do with sex. Finally, kinship, the culturally ostensible "point" of sexual reproduction, is by not means assured; as Haraway (1989) argues, sexual reproduction in fact precludes kinship.

Moreover, the ambivalence of the fusion between "nature" and "culture" is particularly distilled in discussions of kinship, and motherhood as the exemplar of kinship. As Lock argues " ... nature/culture boundaries are contested, and nature is called upon to do cultural 'work' – that is, it participates in commentary on social life, and it forces itself, selectively, into our consciousness" (1997: 273). On one hand, biology is more than ever inextricably linked with social constructionism. But on the other hand, biology is taken as an ontological reality which exists independent of cultural constructions. The public debates we witness concerning reproductive technologies and xenotransplantation testify to the difficulty of traversing these complex relationships. Biological "facts" are "not as self-evident as they might appear" (Franklin, 2001: 304), and cultural notions of kinship are constantly challenged to keep up with both scientific "discoveries" and lived experience.

Chimerism and mosaicism, as yet another scientific "discovery" that propels questions about culturally established kinship relations, highlights the complexity of relations between nature, science, and culture. As this chapter has sought to illustrate, chimerism, as xenotransplantation, challenges traditional kinship boundaries between human and nonhuman animals. Chimerism and mosaicism, as blood, skin cell, and/or genetic genealogy, dramatically challenge the most powerful and enduring of cultural constructions of kinship relations, that between mother and child. As with new reproductive technologies, science here does not reinforce cultural ideology, but offers instead a remarkable challenge. I am certainly not arguing that chimerism and mosaicism are phenomena well known to the public, nor am I arguing that chimerism and mosaicism pervade all "blood" and genetic relations (although the available research clearly indicates that they are much more common than first thought). But, as uncontested "natural" phenomena, in the crucial sense that they do not involve "human" technology, chimerism and mosaicism do not reify the expected linear relationship among the "natural" world, science, and culture. That is, scientific knowledge of chimerism and mosaicism does not confirm the cultural expectation that traditional boundaries between kin and non-kin will be reified. Brown and Michael point out that "publics and the media are exposed to a version of science that most practicing scientists would not recognize. That is, a science stripped of its nuance, uncertainty and richness" and that " ... it would take a vault of spectacular dimensions to imagine that science might, in fact, benefit if its uncertainties were laid bare to public scrutiny on a more transparent basis" (2001: 280).

In many ways, the uncertainties of scientific understandings of "nature" are realized in the seemingly constant shifts that the public experiences in scientific "discovery" and technological innovation. Developments in reproductive technologies and chimerism (as xenotransplantation and as blood and genetic genealogy), represent, for the public at least, a constant (and often uncomfortable) reminder that cultural notions of kinship are subject to change. I am certainly not arguing that science (or "nature" for that matter) is purposefully at the vanguard of the challenges to cultural kinship ideology. It is worth noting here that biologists are not (at least in the studies I have described here) directly interested in questions of kinship. Moreover, "scientific notions of identity do not necessarily displace the social categories of 'race', class and gender" (Fraser 1999/2000: 56). Nonetheless, these studies are of much interest to social scientists concerned with how "nature" is used within the public imagination. Butler notes that it is "only from a self-consciously denaturalized position [that] we can see how the appearance of naturalness is itself constituted" (1990: 110). Kinship, like an increasing number of cultural constructs, may indeed, as Schneider argues, become whatever new facts about biogenetic relationships say it is, but these "new facts" will always be filtered through a set of powerful cultural discourses.

Suggested readings

Daly, M. (1978) "The Cost of Mating," *American Naturalist*, 112: 771–4.
Margulis, L. and Sagan, D. (1986) *Origins of Sex*. New Haven, CT: Yale University Press.
Margulis, L. and Sagan, D. (1997) *What is Sex?* New York: Simon and Schuster.

6
Sex Diversity in Nonhuman Animals

> The universe is not only queerer than we suppose, it is queerer than we can suppose.
>
> (J.B.S. Haldane, 1928: 298)

> When animals do something that we like we call it natural. When they do something that we don't like, we call it animalistic.
>
> (J.D. Weinrich, 1982: 203)

Introduction

When I was a child, Sunday night family television viewing always included Walt Disney. On some occasions the program would focus on a "family" of (usually) bears (but sometimes big cats or dolphins), and tell a story about the youngest cub getting temporarily lost, learning how to fish, or some such life lesson. Even as a child I found it incredible that these animals' lives seemed to mirror human lives so completely, even though the cubs never went to school, had chores, read or wrote, and so on. As an adult, my skepticism toward the ways in which nonhuman animals supposedly exemplify human animal qualities like "the" family, fidelity, selfless care for young, and, perhaps above all, sex complementarity, has only increased.

One of the things I continue to find particularly fascinating, and troubling, is the way in which nonhuman animals are burdened with the task of assuming supposedly universal human qualities. And my childhood suspicion that this "order" should be reversed has been borne out through research. Chapter 4 details Manual De Landa's notion of "organic chauvinism," which usefully suggests that the "point" of life is not human animals – life existed before humans and will continue after humans become extinct – and further, that the "point" of life is not to emulate humans. That is, rather than nonhuman animals mimicking supposedly human qualities, it is far more likely that human animals resemble nonhuman animals.

In most cultures, and for most people, nonhuman animals are symbolic. It matters less how nonhuman animals behave, and more how we think they behave. As the quotes at the beginning of this chapter allude, nonhuman animals serve to confirm our assumptions about the "nature of things" and human beings' relationship to this "nature." The meanings we ascribe to nonhuman animals may indeed have very little to do with the biological and social realities of non-human animals (Bagemihl, 1999).

The aim of this chapter is to explore the ways in which nonhuman animals express a vast diversity of sex, sexuality, kinship, reproduction, and so on, found in the large majority of life on this planet. Chapter 7 will explore this same sex diversity in human animals. The chapter will challenge the range of assumptions made about the "nature" of sex, gender, and sexuality through the use of research on nonhuman living organisms. To do this, I will draw upon what Bruce Bagemihl (1999) refers to as the "quiet revolution" in biology: the overturning of a number of fundamental concepts and theories in evolutionary theory and biology more generally that have anchored traditional (and erroneous) ideas such as the absence of homosexuality in nonhuman animals. To begin, I want to consider "culture" as one of the most common, and therefore powerful, paradoxes concerning the representation of nonhuman animals.

Plants and nonhuman animals are constantly discussed in human cultural terms. In other words, the supposed "laws of nature" are continually read through the lens of social relations (Bagemihl, 1999). The use of terms familiar within human animal cultures to describe behaviors in nonhuman organisms is widespread in biological research. Terms such as "rape," "coy," "cuckoldry," "adultery," "harem," and "homosexual" are extensively used in sociobiology. For instance, scientists use rational choice models to discuss "divorce" in birds (birds will "divorce" if the benefits of selecting an alternate mate outweigh the costs) (Choudhury, 1995; Milius, 1998). Denniston refers to female chaetopod annelids as "wives" (1980: 28) and Ridley has it that female and male gorillas "marry" each other (2003: 20).

Patricia Gowaty strongly objects to the use of these terms on the grounds that they have potentially damaging social repercussions, are more emotionally evocative, and their use is often sensationalized (1982: 630). Gowaty illustrates by using the example of the term "rape" in Thornhill's (1980) study of "Rape in *Panorpa* Scorpionflies and a General Rape Hypothesis." Thornhill defines rape as the apparent forced insemination or fertilization that enhances male fitness and decreases female fitness (fitness as defined entirely in sexually reproductive terms). Gowaty (1982: 630) was asked at a scientific meeting "is it rape when a virgin is forced to intercourse?" and "is it rape when a post-menopausal woman is forced to intercourse?" In human terms, both of these instances are clear acts of rape. However, in Thornhill's sociobiological terms, neither is rape because the virgin's fitness may be increased if she becomes pregnant, and the fitness of the male who copulates with the menopausal woman could not increase, nor is the fitness of the menopausal woman decreased. Other words, like "divorce," also lead to incorrect assumptions. Susan Milius (1998) reviews research on "divorce" in birds, acknowledging that only a minority of birds dissolve pair bonds – because most birds do *not* form pair bonds in the first place.

This is not to suggest, however, that nonhuman animals do not have cultures. Culture may be defined as that "complex whole which includes knowledge, belief, art, morals, custom, and many other capabilities and habits acquired by man [sic] as a member of society" (Taylor, 1871/1927). The history of philosophy, anthropology, and zoology is replete with analyses bent on defining human uniqueness in nature. Aristotle defined our supposed uniqueness as our politics, Descartes as our ability to reason, and Marx by our ability to make conscious choices. Whether it is reasoning, having sex for pleasure, making tools or waging war, nonhuman animals have challenged the assumption that only humans have these abilities (Ridley, 2003).

Thus, although most anthropologists limit culture to the distinctly human realm, studies reveal that nonhuman animals are indeed cultured. Take traditions, for example. Tradition is a learned behavior complex that appears over generations in a shared population. Examples include tool-making and play. Traditions have been documented not only in primates, but birds as well have traditions of foraging, communication, social organization, predator recognition, home, homing and migration, and tool-using (Mundinger, 1980). Moreover, the use of symbolism is now known to be used by nonhuman species.

As another example, language is often taken as the quintessential example of culture. While the ability to learn a language is a capacity of humans with the appropriate neural instructions, what language(s) an individual actually learns will be determined by the environment: hence

English children learn English, and Japanese children learn Japanese. Bird song also fits this definition of culture. Birds have the necessary neural instructions to form various song iterations, but are dependent upon their specific environment which determines the song language they will learn and use. That is, an infant bird *learns* specific song patterns depending upon the bird noises that an infant bird grows up hearing, and these bird songs are *transmitted* over generations by *imitation*. So while we might say that human and bird cultures are different, it is no longer feasible to maintain that only humans have culture. The point here is that there are not simply two cultures, human and nonhuman. There are as many cultures as there are species with cultural behavior because each species is neurophysiologically unique. Language is also interesting because it emphasizes individual variation as central to the concept of culture, a feature often overlooked by anthropologists (Mundinger, 1980). Just as variation is central to natural selection, so too is variation central to cultural evolution.

The point to be made here is that care must be taken in making assumptions about nonhuman living organisms. On one hand, we seem to be keenly adept at projecting human cultural practices such as marriage and divorce onto nonhuman animal behavior. On the other hand, we seem equally reluctant to attribute other signifiers of culture such as language patterns and the use of symbolism. According to Margulis and Sagan, this reluctance stems from a humanocentric desire to ensure not just the uniqueness but the superiority of humans over all other living organisms:

> ... the sheer number of traits listed to explain human uniqueness is enough to arouse suspicion. Among the dazzling array of reasons implying our superiority over the rest of life, one scientific argument stands out to us in curious contrast to the rest: humans are the only beings capable of wholesale self-deception. (1997: 221)

Sex complementarity

The importation of sex complementarity onto nonhuman animals is one of the most powerful examples of the projection of human cultural values onto other living organisms. The history of plant and animal research reveals a legacy of sex complementarity, or the transplantation of sex roles onto plants and animals. Londa Schiebinger (1993) provides a careful historical analysis of the ways in which common, taken-for-granted taxonomic classifications of both plants and animals were actually founded on a political desire to differentiate between human women and men's roles in society. Recalling the point made in Chapter 2, the Enlightenment project which established science as the hegemonic purveyor of knowledge, was girded not only by the creation of sex complementarity, but that this complementarity was produced through "nature."

Take plants, for example. Most flowers are intersex, having both "female" and "male" sex organs in the same individual plant. As we saw in Chapter 2, the move toward gender complementarity in the eighteenth century preoccupied scientists with a desire to validate sex dimorphism. Botanist Robert Thornton emphasized morphological features of the Linnaean taxonomy for plants and animals, features that were *least* important for sexual reproduction, but that emphasized differences between females and males (Schiebinger, 1993). Stamens became defined as "male," and pistils were defined as "female." Botanist Rene-Lois Desfontaines then concluded that stamens had visible orgasms while pistils showed little sexual excitement, "as if the law requiring a certain modesty of females were common to all organisms" (in Schiebinger, 1993). Rather than acknowledge that plants are intersex, botanists like Robert Thornton chose instead to transpose human cultural values onto plant life. The word "*gamete*," referring to the ability of a germ cell to fuse with another cell to create a new individual cell, is derived from the Greek term *gamein*, meaning "to marry." Thus, plant sexuality was confined to the human cultural value of marriage, and by implication, monogamy. Indeed, William Smellie, the chief compiler of the first edition of the *Encyclopaedia Britannica*, felt that pollen, flying "promiscuously" aloft, would produce universal anarchy and cover the earth with "monstrous productions" (in Schiebinger, 1993: 30). By emphasizing the sanctified "marriage" of plants, botanists sought to curb any representations of pollination as a "natural" form of sexual reproduction outside the bounds of monogamy.

Bees were also seconded to endorse the emerging discourse of sex complementarity. John Merrick (1988) illustrates that bees were highly symbolic within European monarchies during the Middle Ages, for instance in adorning the ermine robes of Napoleon. Bees represented strength of spirit and body, and a perfectly organized society based on a strict hierarchy of sovereign bee with complete power over his worker bees. The only problem was that the head of any bee colony is female. Schiebinger remarks that "social function – the act of wielding sovereignty – held greater sway in assigning sex than did the biological act of giving birth" (1993: 23). Thus, female bees became male bees, in a cultural transsex maneuver.

A final example brings sex complementarity closer to our human home, to consider why mammals are called mammals. The term *"mammalia"* means "of the breast," and is another term used in the Linnean system of taxonomy. Given that milk-producing breasts (mammae) are only one characteristic of many that could have been used to classify this group of animals, and further that only half of this group of animals' breasts (i.e. females) are theoretically capable of producing milk, and further that those females that actually do produce milk will only do so for a very short period of time, proved less important to biologists than reinforcing the emerging emphasis on sex complementarity. According to Londa Schiebinger (1993), humans were classified as mammals in order to reinforce a particular notion of femininity, one that was specifically related to the care of infants and children. At the time that the Linnean taxonomy was introduced, the vast majority (up to 90 percent) of middle- and upper-class families sent their children to wet nurses for up to four years (Schiebinger, 1993). The widespread use of wet-nurses had increased the mortality rates of middle- and upper-class infants and children. The Enlightenment project would depend upon a separation of public and private spheres, which meant that women needed to be encouraged to care for their own children. Combined with these factors was the birth of modern medicine and what Michel Foucault (1980) terms "technologies or regulation" and "techniques of self" whereby married couples were incited to monitor, at a microlevel, the behaviors of their children, and the family in general. Thus, politicians, health care practitioners and the clergy conjoined their efforts to both naturalize and normalize what has now become known as the modern family; a system crucially based upon women's roles as "natural" caregivers. At the same time, conceptions of masculinity changed. The term *"Homo Sapien,"* meaning "man of wisdom" separated humans from nature. This contrasted sharply with the classification of humans with other animals as mammals. This contradiction was sustained through sex complementarity, specifically by associating masculinity with reason and femininity with brutish nature (Schiebinger, 1993).

The historical legacy of sex complementarity is found in contemporary research on animal behavior. As these brief examples illustrate, one of the most important ways in which researchers have reinforced sex complementarity is through the *a priori* classification of physical traits and behaviors as *either* female or male. For instance, Richard Estes (1991) theorizes that female bovids have evolved horns because they are "mimicking" male secondary sex characteristics in order to protect their male offspring from dominant male attacks. In other words, female bovids grow horns out of maternal protectiveness (recall from Chapter 4 that this is not how evolution works). Here is one of the essential problems with sex complementarity: females and males are assigned distinct traits or characters. Then when animals exhibit the traits of the "opposite" sex, these traits remain sex-segregated, even in the face of bald evidence that the traits are not sex exclusive. So horns in bovids are defined as a "male characteristic" even when females also grow horns. Indeed Estes relates that only one-third of female bovids do not have horns, which means that two-thirds of female bovids do have horns, making the classification of horns as "male" even more curious. In another example, Katherine Muma and Patrick Weatherhead (1989) describe "male plumage characteristics" in female red-winged blackbirds, concluding that this expression has "no functional value (1989: 23)".

Some feminist researchers argue that sex complementarity is achieved primarily through the silencing of female agency. Sarah Hrdy (1986) and Patricia Gowaty (1997a, c) make the point that scientists tend to study things they are interested in finding out that have a direct or indirect bearing on their own lives and experiences. That is, "empathic responses" determine what scientists will study. Hrdy and Gowaty suggest that much of the research on nonhuman animals has been conducted by male scientists, with an interest in examining facets of animal lives that interest men. Gowaty outlines several principles about the study of female ornithology. The first principle is that females are interesting in their own right, and not just as appendages to males. Incredible as it sounds, females of many species of nonhuman and human animals have only been sporadically studied. For instance, the US legislature has only recently made the inclusion of women in National Institutes of Health research compulsory in epidemiological, disease and treatment trials. This failure to include females in studies produces a related problem which is the inability to understand intrasex differences. That is, when female behavior is mentioned, it is usually assumed that all females of a particular species behave in the same, or similar, way. This creates a paucity of information about how important differences between females creates different outcomes (for instance, in procreation, parental care, etc.). The third principle is that most females seek sex at least sometimes, challenging the myth that females are passive with regard to sexual relations. Even something as seemingly mundane as active female sexuality has only relatively recently been seriously considered within the animal biology literature. The old story that males (nonhuman and human) are sexually assertive with all females compared with females' (complimentary) propensity to choose sexual mates selectively, is still widely propounded today, especially in popular culture. A wealth of research, however, has begun to focus on the ways in which females are very active in their sexual behavior. Researchers are now conceding that female choice of male sex partners is widespread (Birkhead and Møller, 1993b. See also Shaw and Darling, 1985). Female control of paternity may take place at any one or more of several stages: before copulation; during copulation; after copulation but before fertilization; and/or following fertilization. For instance, female birds may be able to determine whether ejaculation occurs, and if it does, the number of sperm that are transferred. Females might also control paternity through the timing of copulation. Again in birds, most females stop copulating well before the end of their fertile period. Thus if a better quality male becomes available, the female can use his sperm to fertilize the majority of her eggs – if a better quality male is not available, she can still fertilize her eggs using the available sperm.

This point relates to the fourth principle, that females are the architects of sperm competition. What Gowaty calls research from a male perspective has historically focused on sperm competition as the sole purview of males. Other research acknowledges the active part that females play in sperm competition. For instance, Birkhead, Møller, and Sutherland (1993) note that the female reproductive tract of mammals and birds is particularly hostile to sperm, with only a very tiny proportion of sperm ever reaching the ova. Some birds, such as female boll weevils, control sperm movement in their reproductive tracts. In birds generally, most of the sperm transferred during copulation is retained by the female, the rest being digested (as a source of nutrients), ejected, or destroyed. Several explanations have been hypothesized, including the avoidance of infection, unfit sperm, and polyspermy. The authors suggest another explanation – that female mammals and birds routinely have sex with numerous males in order to allow the sperm of different males to compete for fertilization. Of course, following evolutionary principles, the unreceptive reproductive tract might have evolved for other reasons, and females simply take advantage of this adaptation, or female sexual behavior and hostile reproductive tracts may have co-evolved. In a related study, Birkhead and Møller (1993a) note that sperm storage, delayed implantation, and delayed development separate copulation from fertilization, and hypothesize that sperm storage (as well as an unreceptive reproductive tract) encourages females to have sex with other males and encourage sperm competition. Female Hymenoptera control the sex of each offspring because males originate from unfertilized eggs, whereas females come from fertilized eggs. Females thus "decide" whether to release sperm on to the eggs or not (Wilson, 2000). Females of several species of rodents have the ability to abort their fetuses through preimplantational pregnancy blocking or through fetus termination after implantation. Several researchers (for instance Labov, 1981) have hypothesized why females in the presence of unfamiliar males will sometimes abort their fetuses, but the reasons remain unclear.

Other principles include the idea that males (and not just females) choose mates – that is, males as well as females, are "choosy,"[41] that females can and do take care of themselves and their young without male assistance (the "female dependence on male" myth), that females compete and are violent toward each other, and that females and males compete against each other. Moreover, female rhesus monkeys have sex with preferred partners regardless of whether or not the female is ovulating (Snowdon, 1997: 283). It is becoming increasingly clear that males of many species wait for females to initiate sex.

The myth that males are aggressive and females are passive remains prevalent, both in popular culture, the media, and zoological research. Of course, female aggression and violence are prevalent among nonhuman (and human) animals. Take the spotted sandpiper for instance. Ornithologist Lewis Oring has witnessed female sandpipers puncture another female's eye or break her leg. Male sandpipers sit to one side during these fights: "it's in the males' best interest to have the best females" (in Milius, 1998: 154). Among spotted hyenas, females dominate males. In the patas monkey species, females dominate males, and females use special solicitation displays to help males get over their fear of females during mating season. Similarly, pygmy marmoset females let males know they are ovulating by becoming less aggressive toward the males (Snowdon, 1997). Snowdon writes:

> There are no consistent patterns of aggression within or between sexes. Females of many species are capable of initiating injurious aggression, and males in many species appear to be pacific. Males will wait on queue for mating opportunities, or will require considerable sexual solicitation or initiation by the female, not only in monogamous and monomorphic species, but also in some highly polygynous and dimorphic species. Males of many species do not appear to be sex machines driven by hormones or pheromones to initiate and force copulations on females, but males are exquisitely sensitive to the

subtle cues provided by females. Some males show high degrees of mate fidelity and appear to work to maintain a relationship with a particular female. Given all these results, we need to be extremely careful with our assumptions about the role that hormones, aggression, and motivation for infidelity play in human behavior. More than likely, the great diversity reported across species in nonhuman primates will be found within the human species. (Snowdon, 1997: 284–5)

Sex diversity

One of the major aims of this book is to both highlight and challenge the common view that "sex" involves two (and only two) distinct (and "opposite") objects (female and male) and further, that these two sexes behaviorally complement each other. Nonhuman animals are often used to support these assumptions. However, these assumptions are not born out by evidence from studies of either human or nonhuman sex. "Sex" is much more diverse than comprising simply two parts. Although the final chapter of this book will examine what Sadie Plant refers to in the quote at the beginning of Chapter 5 as the "thousand tiny sexes" of bacteria, suffice it to mention here that bacteria completely defy the human concept of "sex." Virtually all plant, and many animal species are intersex. That is, living organisms are often both sexes simultaneously – which means that there are not really "two sexes" at all. Many animal species routinely practice transsex, by changing from one sex to another, either once or several times. Other animals practice transvestism by visually, chemically, or behaviorally resembling the "opposite" sex. And over 4000 known species are parthenogenic; that is, all the individuals are female and they reproduce without sex – what we humans term "virgin birth." The following descriptions provide a few examples of intersex, transsex, and transvestism among nonhuman living organisms.

Intersex

We have already noted that plants are intersex, and that most fungi have thousands of sexes. C. Lavett Smith (1967) outlines various forms of intersex in fish. "Normal hermaphroditism" means that nearly all individuals in a species are intersex; "synchronous hermaphrodites" are species in which female and male sex cells "ripen" at the same time regardless of whether or not the individual can self-fertilize; "protogynous hermaphrodites" function first as females and then transform into males; and "protandrous hermaphrodites" function first as males and then transform into females. Robert Warner (1975, 1984) also documents intersex in fish, arguing that the ability to change sex is a positive adaptive strategy. Warner notes that new reports suggest that intersex in fish is much more common than once assumed. Like other biologists, Warner is concerned to understand why more species do *not* have intersex abilities (for a comprehensive study of intersex in animals see Reinboth, 1975).

Transsex

Sex change has long been known to exist among plants (most of which are intersex already) and aquatic animals. We need to be clear here that we are not referring to some sort of sex "role" change, but complete physical sex change. David Policansky (1982) documents some of the widely distributed, both geographically and taxonomically, sex changing species. Sex change here refers to an organism that functions as one sex during one breeding season and the "other" sex during another breeding season. This definition excludes those organisms that can change sex within one breeding season. Given the selective and reproductive advantages of changing sex, Policansky is interested in exploring why more species do *not* change sex, rather than to explain why some species do have this ability. In other words, in some families of fish, transsex is so much the norm that biologists have created a term for those "unusual" fish that do *not* change sex – *gonochoristic*.

The coral goby, for instance, changes sex both ways, between female and male, depending on a number of circumstances. Among fish, sex change is hypothesized to take place when the relative reproductive value between female and male sexes varies with size (called the *size-advantage model*). That is, when it is better (reproductively speaking) to be female, male fish will change into females and vice versa. When goby fish are placed together, the smaller of two males usually changes sex to become female, and the larger of two females usually changes sex to become male (Nakashima, Kuwamura, and Yogo, 1995). Goby fish will also change sex rather than travel long distances to find an "opposite" sex mating partner.

As a further example, earthworms and marine snails are male when young and female when they grow older. Chaetopod annelids show a similar development, but in certain environmental circumstances will change back into males. For instance, when two females are confined together, one female may kill the other female by biting her in half or eating all the available food. When this female has had sex with a male, the male might then turn into a female and bite her in two (Denniston, 1980).

Transvestism

Researchers have found transvestism to be widespread in the nonhuman animal world. Sometimes transvestism takes a physical form, when animals physically resemble the "opposite" sex. Transvestism might also be behavioral, when a nonhuman animal behaves in ways associated with the "opposite" sex of their species. Some entomologists, for instance, describe transvestism in various insect species. Denis Owen (1988) describes female *Papilio phorcas* (a type of butterfly) who take on "male pattern" wings of other male butterflies that fly faster and are better able to avoid prey.

Bruce Bagemihl (1999) notes that transvestism does not mean taking on activities or behaviors that are considered to be either typically "female" or "male." For instance, the sexual reproduction of offspring is typically considered to be in the female domain. But for sea horses and pipe fish, the male bears and gives birth to offspring. So male sea horses and male pipe fish are not practicing transvestism when they produce offspring. Bagemihl (1999) notes this is also the case for behaviors involved in what biologists term "courtship." In many species, females are more aggressive than males in these behaviors. Should a female in these species behave passively, she would be practicing transvestism.

It is worth noting here that nonhuman animals who engage in transvestite behavior, like their human counterparts, specifically avoid homosexual behavior. The misconception that transvestites (usually male) attempt to be "feminine" in order to attract sexual relationships with men is as erroneous for the nonhuman 'animal world as it is for the human animal world'.

Sexual diversity

The living world does not only express a plethora of sex diversity: this section aims to highlight the sheer variability of sexual behaviors within living organisms, which often sits uncomfortably beside cultural notions of sexual behavior, that tend to emphasize the "normality" of heterosexuality, monogamy, and intraspecies sexuality. Because sexual diversity is so common, I argue that the "challenge is to explain the origins of the obvious placticity in sexual expression, not the origins of any one form of expression, such as same-sex interactions" (Pavelka, 1995: 24).

Family values

There is not enough space in this book to document the almost count-less instances of nonhuman living organisms paraded before the media to testify as to the "naturalness" of conservative human "family values." These values consist of monogamous pair-bonding between opposite-sex partners recognized by society (marriage), sexual behavior for the purpose of reproduction (this precludes sex for pleasure and sex between same-sex partners and partners of different species – except in rare circumstances), the prohibition against incest, parental (and especially female) care of young, and celibacy in nonmarried individuals.

Yet the observation of nonhuman animal behavior reveals these val-ues to be distinctly human. Nonhuman animals engage in a wide range of sexual behaviors, only some of which would be recognized within a "family values" rubric. With regard to monogamy, for instance, cultural norms heavily influence biological data.[42] *Monogamy* refers to having only one mate, usually through one or more breeding season. Monogamy also implies that each pair will only mate with each other, although this characteristic has yet to be proven to exist for most birds and mammals. Indeed, leading biologist Edward Wilson notes that "monogamy, and especially monogamy outside the breeding season, is the rare exception. Parent–offspring bonds usually last only to the weaning period and are then often terminated by a period of conflict" (2000: 315. For a discus-sion of child abuse within nonhuman primates see Reite and Caine, 1983). Since female mammals are physiologically capable of caring for offspring alone, there is no necessary reason for monogamy to be favored among mammals. For those species that do exhibit monoga-mous behavior this does not imply anything about the frequency of sex-ual or social interactions between mates (Kleinman, 1977). That many human cultures espouse a clear preference for heterosexual monoga-mous relationships implies a heavy cultural influence. As Kleinman notes:

> ... the legal, legislative, and executive branches of [US] governments, as well as religious establishments, have generally supported ... sexual dimorphism. Since both politics and religion are dominated by men in Western society, *the result is an apparently monogamous sys-tem, with behavioral correlates, however, that are more synonymous to polygyny. In fact, polygyny commonly occurs.* (1977: 61, my emphasis)

Thus, single parenting, or indeed no parental investment at all, is the *norm* in the nonhuman living world (only 3 to 5 percent of mammals form lifetime heterosexual pair bonds). Yet, in human cultures, single parenting is seen as the antithesis of the "natural" order of things. Among nonhuman living organisms, daycare, fostering, and adoption are common; as are infanticide (many parents eat their children) and incest. To take one example, in a study of spotted sandpipers, Oring *et al.* (1992) found that fully half of the broods had been produced by more than two birds, and thus had a complex parental origin.

Nor do many animals have sex solely or primarily in order to reproduce. There is a general lack of acknowledgment of pleasure as an organizing force in relations between nonhuman animals and evolutionary theory generally. E.O. Wilson (2000) notes that male house flies remain copulating with female house flies for a full hour after all of its sperm are transferred, despite the fact that this prolonged copulation decreases his ability to have sex with other flies (and thus produce more offspring). Wilson notes that some insects have sex for an entire day. In "Feminism and Behavioral Evolution: A Taxonomy" Anne Fausto-Sterling (1997) critiques a number of studies of animal behavior in order to demonstrate how selective behaviors are taken up and reinforced as common (and thus, normative) while other behaviors initially observed seem to vanish altogether in subsequent studies. For instance, Darling recounts masturbatory behavior in stags:

> He may masturbate several times during the day. I have seen a stag do this three times in the morning at approximately hourly intervals, even when he has had a harem of hinds. This act is accomplished by lowering the head and gently drawing the tips of the antlers to and fro through the herbage. Erection and extrusion of the penis ... follow in five to seven seconds ... Ejaculation follows about five seconds later. (in Fausto-Sterling, 1997: 51)

The primary reason, according to Fausto-Sterling, that behaviors such as masturbation are often not reported is because animal behaviorists operating within a traditional evolutionary paradigm focus on reproduction at the population level, rather than individual behavior. A focus on reproduction produces a skewed vision of animal life not only as exclusively heterosexual, but in a narrow "functionalist" sense of sexual activity for the purposes of sexual reproduction only. Thus, sexual activities such as masturbation become difficult to account for in such paradigmatic constraints.

To reiterate, many female animals engage in sex when they are already pregnant, and many animals masturbate. Birth control is not restricted to humans; many animals practice forms of birth control through vaginal plugs, defecation, abortion through the ingestion of certain plants, ejection of sperm, and, in the case of chimpanzees, nipple stimulation. Embryos are also known to kill each other before birth.

Perhaps the single most popular debate about sexual diversity, however, is whether or not homosexual behavior is "natural" or "unnatural." This debate usually pivots on whether or not homosexuality is found among nonhuman animals. To even begin to debate this issue we need to first consider whether the term "homosexual" should be applied to nonhuman animals, which brings us back to arguments about culture and nature. As this chapter has so far argued, culture is not restricted to human animals. So if homosexuality is associated with culture, we should not necessarily occlude homosexuality as a cultural behavior among nonhuman animals. If homosexuality is associated with nature, then explorations of homosexuality within the nonhuman living world make sense too. The fact that homosexuality has been documented (despite strong biases toward the nonreporting of homosexuality) in a vast array of nonhuman animals (see the list at the end of this chapter) has led to serious challenges to sociobiological theory (Bagemihl, 1999). Indeed, homosexual behavior in animals strikes at the heart of traditional sociobiological assumptions that the purpose of sex is reproduction.

Homosexuality

> There is a long and sordid history of statements of human uniqueness. Over the years, I have read that humans are the only creatures that laugh, that kill other members of their own species, that kill without need for food, that have continuous female sexual receptivity, that lie, that exhibit female orgasm, or that kill their own young. Every one of these never-never-land statements is now known to be false. To this list must now be added the statement that humans are the only species that exhibit 'true' homosexuality. Does anyone ever state that we alone exhibit true heterosexuality? (Weinrich, 1982: 207)

Research on homosexuality within the human animal population reveals the enormous diversity of this behavior. Although most debates about the "nature" of homosexuality center on whether homosexuality is a product of culture or biology, in actuality, both must play a part. As Bagemihl argues:

> On the one hand it is no longer possible to attribute the diversity of human homosexual expression solely to the influence of culture or

history, since such diversity may in fact be part of our biological endowment, an inherent capacity for 'sexual plasticity' that is shared with many other species. On the other hand, it is equally meaningful to speak of the 'culture' of homosexuality in animals, since the extent and range of variation that is found (between individuals or populations or species) exceeds that provided by genetic programming and begins to enter the realm of individual habits, learned behaviors, and even community-wide 'traditions'. (1999: 45)

Paul Vasey (1995) contends that academic interest in homosexuality among nonhuman primates only began at the beginning of the twentieth century. Initially, homosexuality was characterized as a product of captivity. Then, as the popularity of hormone research increased, studies began to focus on homosexuality as a product of "abnormal" hormone development.[43] Interestingly, Vasey notes that it was the emergence of sociobiology that really activated interest in homosexuality as a product of evolutionary processes. Recent research on homosexuality among nonhuman animals reveals a number of noteworthy points. First, homosexuality is certainly part of our evolutionary heritage: it can be traced back at least to the Oligocene period, to at least 24–37 million years ago (Vasey, 1995). Lifetime pair-bonding of homosexual couples is not prevalent in mammal species, but nor is heterosexual lifetime pair-bonding. Homosexual behavior occurs in over 450 different species of animals, and is found in every geographic region of the world and in every major animal group (Bagemihl, 1999). Research also finds, not surprisingly, that homosexual behavior in animals does not take any one form, but is enormously diverse, and in some species is more diverse than heterosexual behavior (Pavelka, 1995). Some acts are common in same and "opposite" sex behavior, such as anal stimulation. Bisexuality is widespread as well, with more than half of mammals and bird species engaging in both heterosexual and homosexual activities. Studies also suggest that nonhuman animal homosexual behavior varies in frequency within and between species from nonexistence (i.e. it has not been observed by zoologists) to levels that meet or surpass heterosexual behavior. Homosexual behavior has also been observed in all age groups. There is no evidence that homosexual interactions are more common among males than females. Finally, animals clearly learn sexual behaviors within their social groups and pass sexual behaviors down from generation to generation.

Despite these findings, the bulk of animal research *that has admitted to* documenting homosexual activity has attempted to account for homosexuality, usually using a combination of evolutionary (genetic) and behavior adaptation theories. There are three basic "explanations" for the observation of homosexuality in nonhuman animals. Some researchers argue that homosexual behavior is adaptive (McKnight, 1997; Ruse, 1988); others claim it is maladaptive (Gallup and Suarez, 1983), and still others maintain that homosexual behavior is simply neutral (Futuyama and Risch, 1984).

Ignoring homosexual activity

Historically, homosexual activity has most often been ignored in the ethological literature. This has been achieved in a number of ways. First, sexual activity is assumed to be heterosexual unless absolutely proven otherwise. Researchers often mistake the sex of the animals they are observing. Researchers often commonly exclude homosexual activities from what "counts" as sexual activity. Sometimes this means that sexual activity is only counted if there are males and females present, and other times the definition of sexual activity has been restricted to the "insertion of the penis into the vagina" (Bagemihl, 1999: 96).

Indeed, in her review of the literature on homosexual behavior in nonhuman animals, Anne Dagg (1984) notes how difficult it was to produce the review because researchers referred so obliquely to homosexual acts.[44] In another example, David Abbott (1987) explains different reproductive "success" (i.e. the ability to become pregnant and maintain a pregnancy) in a group of female marmoset monkeys. He defines female animals who assist other female animals in the rearing of offspring as "helpers" (1987: 457), and goes on to speculate that dominate females may suppress the fertility of subordinate females for "selfish" and "spiteful" reasons (1987: 458; see also Dunbar, 1980), thus obviating any consideration of female pair-bonding to produce and rear offspring.

Other ways of ignoring homosexual activity include simply not including observations of homosexual activity in final published reports. Bagemihl notes that "numerous published studies on copulation activities of animals provide excruciatingly detailed descriptions and statistics on the frequency of mounts, number of ejaculations, duration of penile erections, number of thrusts, timing of estrus cycles, total number of sexual partners and so on and so on ... but don't mention homosexual activities" (1999: 100). Recalling the earlier exploration of the ways in which researchers have upheld the human cultural ideal of family values, we find that some research emphasizes those behaviors that fit this ideal at the expense of findings that challenge it. For instance, monogamy is found in less than 5 percent of animal populations, but is considered "regular" compared with homosexual behavior which is much more common but considered "irregular," "aberrant," or "abnormal." At the same time, the rate of heterosexual non-monogamous sexual activity is downplayed, even though researchers know that high numbers of offspring are the result of nonpair-bonded sex – in some populations more than three quarters of all offspring (verified by DNA testing. See Bagemihl, 1999).

Finally, one of the most common and powerful ways in which homosexual activity is silenced is by the refusal to admit that it is, indeed, homosexual activity. This is usually done by arguing that homosexual activity is actually some other kind of activity. One scientist reviews with embarrassment his attempt to rename homosexual activity as aggression:

> I still cringe at the memory of seeing old D-ram mount S-ram repeatedly ... True to form, and incapable of absorbing this realization at once, I called these actions of the rams *aggressosexual* behavior, for to state that the males had evolved a homosexual society was emotionally beyond me. To conceive of those magnificent beasts as 'queers' – Oh God! I argued for two years that, in [wild mountain] sheep, aggressive and sexual behavior could not be separated ... I never published that drivel and am glad of it ... Eventually I called the spade a spade and admitted that rams lived in essentially a homosexual society. (Geist in Bagemihl, 1999: 107)

Examples of this maneuver are very common in the literature. In his exhaustive research on the subject, Bruce Bagemihl found that "a number of scientists have actually argued that when a female Bonobo wraps her legs around another female, rubbing her own clitoris against her partner's while emitting screams of enjoyment, this is actually 'greeting' behavior, or 'appeasement' behavior, or 'reassurance' behavior, or 'reconciliation' behavior, or 'tension-regulation' behavior, or 'social bonding' behavior, or 'food-exchange' behavior – almost anything besides pleasurable sexual behavior" (1999: 106).[45] Bagemihl found researchers explained away instances of oral sex among male Orang-utans as being nutritively motivated.

Denying homosexuality

If homosexual activity is recognized, it is most often explained as "really being something else." The most popular way of denying homosexuality in nonhuman animals is to define homosexual behavior as *pseudo-heterosexuality*. In other words, one individual of the homosexual pair is assigned the "female" role and the other the "male" role, irrespective of the reciprocity of actual actions, and irrespective of whether or not these animals also engage in heterosexual activity. For instance, Bottlenose Dolphins, Cheetahs, and Grizzly Bears practice same-sex pair-bonding to the exclusion of opposite-sex pairing (Bagemihl, 1999).

James Weinrich (1980) presents an interesting, yet ultimately disappointing, analysis of homosexuality in nonhuman animals. A number of homosexual behaviors are documented, starting with the worm *Moniliformis dubius*. These intestinal parasites arrange themselves such that females feed toward the front of the host's intestinal tract and males feed toward the back. When male worms reach maturity, they move up toward the females, inseminating the females and then cementing their genital tracts, presumably to prevent their semen from leaving the genital tract, and perhaps preventing other males' semen from entering the tract. Some males also have sex with male worms along the way, cementing their genital tracts as well. Weinrich criticizes previous research that defined this behavior as "homosexual rape" and notes that the male worms can distinguish between female and male worms because they only deposit semen in the female worms. Weinrich also describes homosexual behavior in various species of fish, the hanging fly, lizards, salamanders, and mountain sheep. Unfortunately, the author is unable to see these behaviors as anything other than imitations of heterosexual behavior: " ... the homosexuality observed is some kind of variation of the heterosexuality observed in the same species" (1980: 293).

Another way of ignoring homosexual activity is to argue that it is really only a substitute for heterosexuality – this may be more familiar as the "men in prison" argument. The theory is that individuals will engage in homosexual activity in the absence of heterosexual partners. As Bruce Bagemihl points out, this argument accedes a great deal of power to homosexuality – "this is actually an unintentional assertion of the relative strength of the homosexual urge, or correspondingly, the relative weakness of the heterosexual imperative – for the stronghold of heterosexuality must be tenuous indeed if such factors are capable of upsetting the balance" (1999: 134). Of course, this argument also ignores the finding that many animals engage in homosexual activity when there are plenty of "opposite" sex partners available, and that individuals who are unable to find a heterosexual partner are usually also unable to find a homosexual partner as well. Other theories include the idea that animals become homosexual because they have been "contaminated" by homosexuality – in human culture this takes the form of the popular myth of the older man seducing the young boy who would otherwise have been heterosexual. Still other theories argue that animals engage in homosexual sex because they misidentify the sex of their partner. Of course, if animals did "misidentify" the sex of members of their own species, we would expect approximately equal amounts of same and "opposite" sex activity, which has not been found. Similarly, the argument that homosexual activity is a product of captivity is countered by the finding that homosexuality is a more common pattern under free-ranging conditions than captivity (Vasey, 1995). Moreover, human zoo personnel typically only allow animals to "choose" "opposite" sex animals to have sex with. Moreover, Kim Wallen (1995) points out that in many mammal species, sexually reproductive behavior is strongly regulated by ovarian hormones, which actually regulate the female's ability to have sexually reproductive sex. In contrast, for humans and other primate species, hormone conditions and female sexual activity are only very loosely correlated. Thus, humans and other primate species are able to have sex at any time, without hormone stimulation.

Tina Adler (1997) notes that studies often attempt to rationalize homosexual behavior as having some "function" within heteronormativity – for instance, female cows mounting each other in order to signal to bulls that they are receptive to sex, or the observation that some animals are more willing to share food with members of their own sex when they have had sex with them. However, despite all strident attempts to "explain away" (Bagemihl, 1999) homosexual sex as merely servicing heterosexual sex, studies now suggest that homosexual behavior might actually have more to do with sexual gratification than reproduction. This seemingly banal suggestion flies in the face of traditional studies, which, in those rare cases that actually refer to homosexual behavior, incorporate homosexuality into heterosexuality. As primatologist Paul Vasey remarks, "the idea that animals may have sex just because it feels good proves difficult for some people to accept" (in Adler, 1997: 2). In his own research on Japanese macaques, Vasey found that mutual sexual attraction was the motivation for the formation and maintenance of homosexual pairing. Vasey concludes, "I'm not saying Darwin was wrong, but there's room for working on the theory so it can accommodate observations of homosexual behavior" (in Adler, 1997: 3). Moreover, in terms of the inevitable comparisons between nonhuman animals' and human animals' sexual behavior, Vasey contends that human homosexuality may have no "evolutionary or reproductive benefits and that it's just for pleasure also" (in Adler, 1997: 4).

Explaining homosexuality

If researchers are unable to convincingly argue that homosexuality does not exist among nonhuman animals, then the task transforms to one of attempting to explain homosexuality. Historically, heteronormative society has made non-heterosexual behavior difficult to acknowledge, let alone accept. As we have already seen, one of the most powerful modern mechanisms heteronormative society utilizes to maintain its hegemony is the differentiation between "normal" and "abnormal" sex and sexuality. The study of nonhuman animals has not escaped this cultural taxonomy.

Edward O. Wilson is one of the most prominent and well-respected contemporary biologists, and his book *Sociobiology: The New Synthesis* (2000) was voted by officers and fellows of the international Animal Behavior Society "the most important book on animal behavior of all time" (2000: back cover). As one of the strongest proponents of a biological foundation to human social behavior, Wilson nevertheless devotes relatively little attention to homosexuality, or sexual diversity more generally. In *Sociobiology*, Wilson offers a somewhat ambivalent approach to homosexuality. On one hand, he refers to female homosexuality as a "aberrant," "deviant" and "a temporary maladaptation" (2000: 22). On the other hand, Wilson reviews a number of theories that attempt to find a "positive" function for homosexuality while maintaining the traditional evolutionary emphasis on natural selection. We find in Wilson's work a synthesis of the major theories advanced to explain homosexuality in nonhuman animals. For instance, in reviewing studies on the South American leaf fish *Polycentrus schomburgkii*, Wilson notes that "subordinate" males sometimes "imitate" females through color change and behavior in order to try to increase their reproductive success by getting close enough to a female's deposited eggs and "fooling" the resident males. Wilson concludes that if this interpretation is correct, "we have here a case of transvestism evolved to serve heterosexuality!" (2000: 22). Wilson suggests that female hyena penises are actually "pseudo-penises" used to appease male hyenas (2000: 229), that male hamadryas baboon homosexual behavior is "true automimicry" (2000: 230), that homosexual behavior in human and nonhuman animals generally is "altruistic" (2000: 311). The theory suggests that altruism may stem from either the homosexual person her/himself, or the homosexual person's parents who attempt to optimize the reproductive success of their offspring by limiting the number of offspring who actually reproduce, and by encouraging those who do not reproduce (presumably the homosexual children) to assist in the caring of their siblings' offspring. Wilson suggests the social pressure placed on some children to become homosexual "need not be conscious" (2000: 343). Wilson reiterates Hutchinson's (1959) now well-cited theory that homosexual genes possess superior fitness in heterozygous conditions. Using the example of sickle-cell anaemia, Hutchinson argues that the homosexual gene would be maintained in evolution if it is superior in a heterozygous state, in which heterozygous genes mature better and/or produce more offspring. Of course, this theory assumes that there is a gene for homosexuality, a theory far from proven. I will take up this theory of a genetic basis to homosexuality later in the chapter.

Although more recent research focuses on nonheterosexual behavior in the nonhuman animal world, as Anne Dagg (1984) notes, the acceptance of homosexual behavior (let alone female-to-male mounting) as normal is far from universal. In her review of the literature, Dagg provides examples of the ways in which cultural norms have structured scientific interpretation:

> In 1965, for example, Buechner and Schloeth noted that in the kob the 'female sexual displays formed a continuum from male-behaving females to normal females' thus restricting the definition of normal, and Wendt (1965) wrote that 'In the wild, substitute sexual activities [such as homosexuality] are not necessary. And where innate dispositions of this sort exist, natural selection rapidly and effectively prevents their perpetuation' ... In that same year, Beach wrote that mounting behavior by females was typical of most females in many species, and noted the paradoxical conclusion of many authors that 'normal' feminine mating patterns include 'masculine' elements. In 1968, Ewer, in her study of *Sminthopsis crassicaudata*, postulated that two barren females which tried to mount a male and a female may have done so because they were from a weak litter and had some hormonal defect, thus also considering their behavior abnormal. (1984: 156)

Here we see, again, the delineation of heterosexual-only behavior as "normal," with every other kind of sexual behavior designated "abnormal" and in need of explanation. We also find the familiar "feminine"/ "masculine" binary in operation such that sexual assertion ("mounting behavior") is necessarily defined as "masculine" even when most females in a species display this behavior. And, as we explored in Chapter 3, hormones are also called upon to account for this "abnormal" behavior. But while Dagg prefaces her review of studies on homosexuality with an acknowledgment of the biases within the literature toward heterosexuality, she nevertheless repeats the bias by uncritically reviewing traditional "explanations" for homosexuality; namely captivity, domestication, phylogenetic relationships, and social organization. She goes on to repeat the familiar "contexts" within which homosexual behavior in nonhuman animals is supposed to take place: social play, physical contact non-play, aggression, and sexual excitement. Dagg notes that these contexts might well be applied to homosexual behavior in human animals, but insists these have "limited and controversial homologies in human behavior, notably in children's play, in rape (aggression), and in such things as encounter groups" (1984: 179). Nowhere does Dagg concede that these "contexts" may just as accurately be understood as the context for *heterosexual* behavior, whether in human or nonhuman animals.[46] Warren Gadpaille (1980) similarly moves from animals to humans in his search to "explain" homosexuality. Initially conceding that homosexual behavior has been reported in a wide variety of species, Gadpaille goes on to define homosexuality as "biologically deviant" from an evolutionary perspective on the basis that natural selection attempts to foster traits that will increase sexual reproductive success (1980: 354). As we know from Chapter 4, individuals display a wide variety of traits and characteristics that are adaptive, neutral, or maladaptive, and all exist under the rubric of evolution.

Interestingly, Gadpaille further stretches evolutionary theory by arguing that certain social and genetic forces have produced, over generations, "broader sex differences involving intellectual and emotional traits and capacities quite unrelated to species reproduction" (Gadpaille quoting Beach, 1980: 354). Gadpaille notes that "although innate psychological masculine/feminine differences are currently objected to by some, neither data nor logic support the objection" (1980: 354). From there, Gadpaille leaps to the conclusion that these innate and inbred "feminine"/"masculine" traits, if fostered and reinforced by parents and other adults, will decrease the likelihood of homosexuality, and that it is "interference" in early childhood that increases the likelihood of homosexuality. Thus, if left alone, children will become heterosexual because of innate sex "differences": it is only with adult "interference" that children become homosexual. The list of problems with this skewed "logic" is long, but most obvious, is the erroneous (if common) confusion between sexuality and gender roles. Homosexuality is not the same thing as "femininity" in men, nor "masculinity" in women. Equally problematic is the unequal use of biology to argue that homosexual behavior in primates is prevalent (i.e. "natural") but homosexuality in human animals is "sexually deviant (1980: 349)."

Other attempts to explain homosexuality abound. Some researchers propose that homosexuality is maintained in the service of heterosexuality by acting as a regulator of reproductive growth and by providing nonreproducing adults who will help raise offspring (Wilson's argument). Of course, the problem with this argument is that animals that engage in homosexual behavior sexually reproduce (indeed, this line of argument seems to me to be incredibly naïve given the number of homosexual humans who sexually reproduce).

Another explanation argues that young animals engage in homosexual activities as a "practice-run" for heterosexual activities. There is no evidence to support this claim: animals do not need to have heterosexual experiences when young in order to develop competent heterosexual activity as adults. Besides, animals, and especially primates, learn very quickly, so it is unlikely that homosexual behavior is primarily a way of practicing heterosexuality (Vasey, 1995). Yet another explanation suggests that homosexuality is actually a breeding strategy insofar as it attracts heterosexual sex and/or builds group cohesion. Using functionist theory (the problems of functionalist theory applied to evolution are outlined in Chapter 4) Parker and Pearson theorize (1976) that when females mount each other, it "functions" to increase the reproductive success of the female that is doing the mounting by mimicking males which will somehow attract males to have sex with them. This assumes that some females only mount, while others are only mounted. It also assumes that mounted females are acting altruistically to benefit the female mounter. Tyler (1984) suggests the same function, but with opposite reasoning. Tyler suggests that females mount to prevent the mounted females from being mounted by males. Neither of these theories are supported by empirical evidence.

John Kirsch and James Rodman (1982) make a similar argument by suggesting that homosexuality in animals (and, by extension, humans) is a positive evolutionary adaptation whereby homosexual animals do not reproduce themselves, and thus have time and energy to devote to assisting heterosexual animals to raise their young. The authors admit this hypothesis remains at the theoretical level, and state that their major aim is not to provide a justification for homosexual behavior but rather to show that the claim that homosexuality is wrong because it is "unnatural" is false. A similar argument maintains that homosexual behavior serves to reduce tension, reconcile animals in conflict, and build alliances.

Finally, a number of researchers propose that homosexual activity functions to establish and maintain dominance hierarchies. This theory depends on a conservative understanding of sexual behavior. Homosexual activity tends to be entirely defined by mounting behavior, with mounting judged as dominating, and being mounted as submissive. Unsurprisingly, we find this definition within human culture, where men are usually defined as active and women are defined as passive. The evidence from nonhuman animal research finds many cases of socially subordinate animals mounting socially dominant animals.

Challenging explanations of homosexuality

Douglas Futuyma and Stephen Risch (1984) eschew sociobiological theories of homosexuality and challenge that "evolutionary theory provides no guide to morality or ethical progress, nor for appropriate social attitudes toward homosexuality" (1984: 157). These researchers contend that much of the vast scientific literature attempting to "account for" homosexuality is motivated by a generally negative attitude toward homosexuality in Western cultures which leads to research designed to provide clues as to how to prevent homosexuality. But in order for a biological account of homosexuality to have any validity, it must be able to verify its theories, which has proven difficult as the above discussion indicates. For instance, a number of researchers (see, e.g. Ruse, 1981; Weinrich, 1976; Wilson, 1975, 1978) argue that homosexuality is an evolved trait. For a trait to evolve, it must mean that genes "program" individuals to develop homosexuality under suitable environmental circumstances, and that over many generations these genes that code for homosexuality have replaced genes that do not code for homosexuality. We need to be clear about what we mean by genetic coding here. All biological characteristics have a *genetic* basis in the sense that the characteristic could not develop unless the organism did not have this information in the DNA. But all characteristics also have an *environmental* basis as well, in the sense that the organism cannot develop the trait unless it is in the right environmental circumstances.

Futuyama and Risch evaluate the evolutionary theory of homosexuality by considering the following issue: "is homosexual behavior a distinct trait or is it simply one manifestation of a more generalized trait, such as sexual behavior" (1984: 161). Futuyma and Risch consider a number of possible explanations. One possibility is that two sets of genes exist, one for homosexual behavior and one for heterosexual behavior. The other possibility is that one set of genes exist for sexual behavior, and this is expressed as either homosexual or heterosexual, or some combination of the two, depending on the environmental context. Futuyma and Risch consider the available evidence consisting of heterozygous advantage, kin selection, and parental manipulation. *Heterozygous advantage* refers to the hypothesis that people who inherit a gene for heterosexuality from one parent and a gene for homosexuality from the other parent will be more likely to survive and reproduce. There is no evidence that this hypothesis is valid. As Futuyma and Risch point out "showing that a trait 'runs in families' does not prove there is a genetic basis: wealth, religious affiliation, and social attitudes also run in families" (1984: 162). *Kin selection* refers to the hypothesis that gay people help to propagate their genes by helping to care for the offspring of their non-gay siblings. As we have already seen, there are several problems with this hypothesis. First, it assumes that gay people reproduce less than heterosexual people (interestingly, the heterozygous advantage hypothesis is based on the opposite supposition – that people with gay as well as heterosexual genes are *more* likely to reproduce). Second, it ignores much research that finds that people act altruistically toward members of a kinship group because of group bonds rather than genetic commonality. Third, the supposition that people want homosexual genes to be inherited because gay people enjoy privileged status in society makes vast assumptions about the sexual hierarchy in society. Moreover, while it might apply to some mainly non-Western cultures in which gay men occupy positions such as priests or shamans, it completely ignores lesbian women, who rarely occupy privileged positions. Finally, *parental manipulation* hypothesizes that parents actually influence some of their children to be homosexual so that the parents' heterosexual children will (somehow) more successfully reproduce. This, according to Futuyma and Risch, is no more than sociobiological speculation. The authors conclude that "one of the weaknesses of sociobiology is that it sees almost every aspect of behavior as adaptation. But organisms display a wealth of nonadaptive (neutral), and even maladaptive, characteristics" (1984: 166). As we have seen, there is no evidence that homosexual behavior is maladaptive because it does not interfere with animals' reproduction. But nor is homosexual behavior necessarily adaptive. In fact, homosexual behavior appears to be neutral.

When researchers debate homosexuality in terms of its "fitness" (read "goodness" and "normality"), it is important to bear in mind that fitness can only be usefully discussed in terms of an entire genotype. As Dobzhansky explains:

> It cannot be stressed too often that natural selection does not operate with separate 'traits'. Selection favors genotypes. The reproductive success of a genotype is determined by the totality of the traits and qualities which it produces in a given environment. (1956: 340)

At the same time, in discussions about the evolutionary "utility" or "purpose" of homosexuality, naturally selected for behaviors are not directed toward the preservation of the species. In other words, only behaviors that are adapted to the individual, and not the larger community or species, will be supported by natural selection (Deichmann, 1996). We can infer that homosexual behavior coextensively developed with the increase in behavioral flexibility, and the decoupling of sexual behavior from reproduction. Moreover, reproduction itself is often peripheral to animal life. For instance, 90–98 percent of Damaraland mole-rats never reproduce during their lifetime, but do engage in sexual activity. And the Damaraland population is nevertheless maintained (Bagemihl, 1999).

Transspecies sexuality

Despite the fact that homosexuality was removed from the American Psychological Association's classification of "abnormal behavior," homosexuality may or may not still be considered a deviation from the heterosexual "norm." That being said, there is a list of other sexual behaviors that few people question as "abnormal." Sex between different species is one of them. People recognize that sexual intercourse between a horse and donkey might produce an ass, but, on the whole, transspecies sex is considered impossible. But as researchers extend their vision beyond the heterosexual "norm," findings are beginning to emerge to suggest that sexual behavior among nonhuman animals is again much more plastic and diverse than human culture allows. Sexual behavior between flowers and various insects is so commonplace that it is rarely recognized as transspecies sexual activity. But other examples have been found. For instance, Krizek (1992) observed and documented a sexual interaction between two different orders of insects; a butterfly and a rove beetle. The rove beetle was perched on a leaf with its abdomen elevated. The butterfly approached and for several seconds explored the beetle's anogenital organs with its proboscis. Krizek notes that other such interactions, between different orders of human and nonhuman animals, have been observed.

Conclusions

Although less than 2 million species of nonhuman animals are currently classified, researchers estimate the existence of between 5 and 50 million species of nonhuman animals on earth (May, 1988). The diversity of sex and sexual behavior of living organisms on this planet is far more diverse than human cultural notions typically allow. This diversity confronts cultural ideas about "the" family, monogamy, fidelity, parental care, heterosexuality, and perhaps most fundamentally, sexual difference. Research on nonhuman animals immediately raises a number of issues. Nonhuman animals are closely linked with "nature"; thus what animals do is considered to be "natural." In Western cultures, "natural" is often attached to morality – "nature" becomes "natural" becomes "good." So when animals behave in ways that apparently reinforce normative conceptions, the moral currency follows smoothly. Problems occur when nonhuman animals do not behave in ways that are obviously interpretable within the normative framework. As this chapter has argued, in these cases, researchers have often silenced, erased, or variously accounted for these behaviors such that any confrontation with norms are minimized. As Bagemihl notes "naturalness is more a matter of interpretation than facts" (1999: 78). In these cases, the link that researchers and the public usually make between human and nonhuman animals is de-emphasized, and the "uniqueness" of humanity is underscored. In some ways, it is difficult not to assent to differences of kind between human and nonhuman animals. However, this chapter has argued that the major difference is the *attribution of meaning* to nonhuman animal behavior. Homosexuality is a case in point. Human interpretations of homosexuality among nonhuman animals is dependent upon a number of factors. First, to even define certain behaviors among animals as "homosexual" involves the invocation of a naming process with a particular social history. As social scientists, we may well argue that it is impossible to attribute certain motivations to human animals, and this is only compounded by the fact that nonhuman animal behavior is entirely inferred through observation. To discuss homosexuality among animals is to necessarily invoke a particular framework of meaning. When this framework is applied, it is clear that homosexuality is as "natural" as heterosexuality among nonhuman living organisms. By this I mean that homosexuality has been observed in virtually all animal groups and in all areas of the world. The most notable difference between human and nonhuman animal homosexuality is that, among nonhuman animals, homosexuality does not invite negative reactions from other animals:

> By defining the boundary that separates other primates from humans, primatologists mold society's ideas of human nature. Although the first reports of homosexual behavior among primates were published >75 years ago, virtually every major introductory text in primatology fails to even mention its existence. Insofar as nature is often the popular criteria for crafting moral and social policies, one might be left with the impression that homosexual behavior is a recent abnormality unique to humans, and thus outside the natural order. Nevertheless, there exists robust evidence that homosexual behavior, and by extension, other nonreproductive sexual behaviors, are the products of a long evolutionary history that occurred independent of human culture. While homosexual behavior is widespread among our primate relatives, aggression specifically directed toward individuals that engage in it appears to be a uniquely human invention. (Vasey, 1995: 197)

In the nonhuman living world, homosexual activity is treated as routinely as heterosexual behavior. Indeed, Bagemihl makes the point that "in many species it is heterosexual, not homosexual, behaviors that draw a negative response. In numerous primates and other animals, male–female copulations are regularly harassed and interrupted by surrounding animals" (1999: 55). LeVay also concurs that "there seems to be no homophobia in the animal kingdom ... In biological terms, homophobia is deeply incomprehensible" (1996: 209). Another way of looking at this is that heterosexuality is as "unnatural" as homosexuality, insofar as heterosexuality is often accompanied by behaviors that humans condemn such as nonreproduction, promiscuity, hostility, and general instability (recall that only about 3–5 percent of mammals form life-long pair-bonds).

Although homosexuality courts particular debate, we may apply the same "natural"/"unnatural" argument to non-normative sexuality in general. To the extent that evolutionary theory has focused on the saliency of sexual reproduction, we have missed a banal but nonetheless pivotal point: that the existence of something *is* its "function" (Bagemihl, 1999). Just as the enormous diversity in human sexual behavior cannot be shoe-horned into sexual reproduction, neither can nonhuman sexual behavior. Murphy summarizes the conundrum humans have created by claiming a "natural" basis for heteronormativity:

> If nature is understood as the world with human animals in it, then homoeroticism is manifestly natural; if nature is understood as the world without human animals, then it is very difficult to determine what living organisms and what behaviors should count as 'exemplars for human morality'. (Murphy, 1997: 189)

Table 6.1   A sample of Organisms in which Sex/ual Variation has Been Observed (compiled from Bagemihl, 1999)

Primates	*Marine mammals*	Pronghorn
Bonobo Chimpanzee	Boto Dolphin	Kob
Common Chimpanzee	Bottlenose Dolphin	Waterbuck
Gorilla	Spinner Dolphin	Lechwe
White-Handed Gibbon	Orca or Killer Whale	Puku
Siamang	Gray Whale	Blackbuck
Hanuman Langur	Bowhead Whale	Thomson's Gazelle
Nilgiri Langur	Right Whale	Grant's Gazelle
Proboscis Monkey		Bighorn Sheep
Golden Monkey	*Seals and manatees*	Thinhorn or Dall's Sheep
Japanese Macaque	Gray Seal	Asiatic Mouflon or Urial
Rhesus Macaque	Northern Elephant Seal	Musk-Ox
Stumptail Macaque	Harbor Seal	Mountain Goat
Bonnet Macaque	Australian Sea Lion	American Bison
Crab-Eating Macaque	New Zealand Sea Lion	Wisent or European Bison
Pig-Tailed Macaque	Northern Fur Seal	African Buffalo
Crested Black Macaque	Walrus	Mountain Zebra
Savanna Baboon	West Indian Manatee	Plans Zebra
Hamadryas Baboon		Takhi or Przewalski's Horse
Gelada Baboon	*Hoofed mammals*	Warthog
Squirrel Monkey	White-Tailed Deer	Collared Peccary or Javelina
Rufous-Naped Tamarin	Mul or Black-Tailed Deer	Vicuna
Verreaux's Sifaka	Wapiti, Elk or Red Deer	African Elephant
Lesser Bushbaby or	Barasingha or Swamp Deer	Asiatic Elephant
Mohol Galago	Caribou or Reindeer	
	Moose	
Other mammals	Giraffe	
Carnivores	Whiptail or	Least Chipmunk
Lion	Pretty-Faced Wallaby	Olympic Marmot
Cheetah	Rufous Bettong or	Hoary Marmot
Red Fox	Rat Kangaroo	Dwarf Cavy
(Gray) Wolf	Doria's Tree Kangaroo	Cui or Yellow-Toothed Cavy
	Matschie's Tree	Aperea or Wild Cavy

Table 6.1 Continued

Bush Dog	Kangaroo	Long-Eared Hedgehog
Grizzly or Brown Bear	Koala	Gray-Headed Flying Fox
Black Bear	Northern Quoll	Livingstone's Fruit Bat
Spotted Hyena		Vampire Bat
	Rodents, Insectivores,	
Marsupials	*and Bats*	
Eastern Gray Kangaroo	Red Squirrel	
Red-Necked Wallaby	Gray Squirrel	

Waterfowl and Other Aquatic Birds

Greylag Goose	Black Stilt	Red Bishop Bird
Canada Goose	Oystercatcher	Orange Bishop Bird
Snow Goose	Golden Plover	House Sparrow
Black Swan	Ring-Billed Gull	Black-Billed Magpie
Mute Swan	Common or Mew Gull	Acorn Woodpecker
Mallard Duck	Western Gull	Anna's Hummingbird
Blue-Winged Teal	Kittiwake	Blue-Bellied Roller
Lesser Scaup Duck	Silver Gull	Pied Kingfisher
Australian Shelduck	Herring Gull	Griffon Vulture
Musk Duck	Black-Headed Gull	Ruffed Grouse
Common Murre or	Laughing Gull	Sage Grouse
Guillemot	Ivory Gull	Kestrel
Laysan Albatross	Caspian Tern	Humboldt Penguin
Great Cormorant	Roseate Tern	King Penguin
European Shag	Guianan Cock-of-the-Rock	Gentoo Penguin
Silvery Grebe	Calfbird	Greater Rhea
Hoary-Headed Grebe	Swallow-Tailed Manakin	Emu
Black-Crowned	Blue-Backed Manakin	Ostrich
Night Heron	Bicolored Antbird	Regent Bowerbird
Cattle Egret	Ocellated Antbird	Superb Lyrebird
Little Egret	Ocher-Bellied Flycatcher	Victoria's Riflebird
Little Blue Heron	Tree Swallow	Raven
Gray Heron	Cliff Swallow	Jackdaw
Pukeko or Purple	Bank Swallow or	Wattled Starling
Swamphen	Sand Martin	Raggiana's Bird
Tasmanian Native Hen	Hooded Warbler	of Paradise
Dusky Moorhen	Chaffinch	Long-Tailed Hermit
Hammerhead	Scottish Crossbill	Hummingbird
Flamingo	Red-Backed Shrike	Black-Rumped Flameback
Ruff	Blue Tit	Galah or Roseate Cockatoo
Buff-Breasted Sandpiper	Eastern Bluebird	Orange-Fronted Parakeet
Greenshank	Gray-Capped	Brown-Headed Cowbird
Redshank	Social Weaver	Peach-Faced Lovebird
Black-Winged Stilt	Sociable Weaver	

Suggested readings

Abramson, P. and Pinkerton, S. (eds) (1995) *Sexual Nature, Sexual Culture*. Chicago, IL: University of Chicago Press.

Bagemihl, B. (1999) *Biological Exuberance. Animal Homosexuality and Natural Diversity*. New York: St. Martin's Press.

Birkhead, T.R. and Møller, A. (1993b) "Female Control of Paternity," *Trends in Ecology and Evolution*, 8: 100–4.

Denniston, R.H. (1980) "Ambisexuality in Animals," in J. Marmor (ed.) *Homosexual Behavior: A Modern Reappraisal*, New York: Basic Books, pp. 35–40.

Futuyama, D.J. and Risch, S.J. (1984) "Sexual Orientation, Sociobiology, and Evolution," *Journal of Homosexuality*, 9: 157–68.

7
Sex Diversity in Human Animals

Do we truly need a true sex? With a persistence that borders on stubbornness, modern Western societies have answered in the affirmative. They have obstinately brought into play this question of a "true sex" in an order of things where one might have imagined that all that counted was the reality of the body and the intensity of its pleasures.

(Foucault, 1980: vii)

We have the power to shake people's foundations. You can't touch one of us and come away unchanged. At least you'll come away questioning things. We infect people, just by being. We are walking carriers of the gender questioning disease.

(Julian in Preves, 2003: 125)

Introduction

Human animals, like other animals, actualize a vast range of kinship practices, sexual practices, sexually reproductive practices, and so on. Cultural mores may prohibit the frank discussion of these practices, but they take place nonetheless. Many books, articles, television, and radio programs are devoted to detailing this diversity of sexual, kinship, and reproductive practices of human animals. But one aspect of human animal diversity that has only recently caught the attention of the public is the diversity of sex "differences." Recognition of this diversity means, at the very least, understanding that there are more categories than simply "female" and "male." The term "*intersex*" (which, before 1920, used to be referred to as "hermaphroditism") refers to human and nonhuman

122

animals who are born with physical characteristics that have been taxonomically defined as *either* female or male.

The aim of this chapter is to introduce intersex as a not uncommon natural source of sex diversity among human animals (see Chapter 6 for a discussion of intersex in other animals). While the chapter reviews the etiology of some of the most common forms of intersex, the major part of the chapter is reserved for an analysis of the changing cultural response to human sex diversity. I will argue that the major shift from a "one-sex" to a "two-sex" model (discussed in Chapter 2) has had major implications for people with intersex conditions. The modern Western treatment of intersex conditions relies heavily on surgery to *physically* conform sex diversity to the heteronormative system of sex comple- mentarity (as outlined in Chapter 1). This surgical reconstruction is taking place within the context of a growing community of people with intersex conditions who contest the nonacceptance of sex diversity in human animals.

In this chapter, I want to suggest that the modern regulatory tech- nique of medical surgery radically disciplines the physical intersex body. As well, it raises serious questions about the possibilities for, and limits of, transgression. As such, the chapter focuses more on societal responses to the intersexed body than on intersex as a subjectivity. Through this analysis, I focus on the ways in which certain develop- ments in medical technologies during the twentieth century led to changes in the way in which the body was encoded by medical science. These particular configurations of knowledge have enabled the deploy- ment of a variety of technologies in the "treatment" of people with intersex conditions, which has led to the creation of new subject posi- tions in the latter part of the twentieth century.

The variability of sex[47]

Intersex conditions among human animals provide a valuable opportu- nity to explore the relationship between "sex" and "gender," as well as the designation of meaningful categories of difference. It is difficult to get a completely accurate estimate of the prevalence or incidence of intersex conditions. Cheryl Chase (1998) estimates that one in every hundred births shows some morphological "anomaly," observable enough in one in every two thousand births to initiate questions about a child's sex. Anne Fausto-Sterling (2000) suggests 2 percent of live births, approximately 80 000 births per year, demonstrate some genital "anomaly". Out of those, approximately 2600 children a year are born with genitals that are not immediately recognizable as female or male. Milton Diamond estimates the incidence slightly lower, at 1.7 percent of the population (2000). Intersex is an umbrella term, under which a variety of conditions are placed, including androgen insensitivity syndrome (AIS), progestin induced virilization, adrenal hyperplasia, Klinefelter syndrome, and congenital adrenal hyperplasia (CAH).[48] The term "intersex" actually refers to a whole range of different conditions that human animals (and, of course, other animals) may be born with. Here are a few examples.

Androgen insensitivity syndrome (AIS) is a genetic condition (although there are cases, as we would expect from nature, of spontaneous mutations) affecting approximately one in 20 000 individuals, in which the body's cells are unable to respond to androgen (www.isna.org/faq/faq-medical.html). Some individuals have complete AIS: here the karyotype is 46XY and the testes produce mullerian inhibiting hormone (MIH) in utero, which results in a fetus without uterus, fallopian tubes, cervix, and upper part of the vagina. But because the cells do not respond to testosterone, the genitals differentiate as female. Thus, at birth, the child will have "normal" looking female genitals, plus undescended or partially descended testes and a short, or sometimes absent, vagina. At puberty the young person will show breast growth, but will not menstruate, will have little or no pubic or underarm hair, and will not be able to sexually reproduce. Because AIS is genetic, it runs in families.

Progestin induced virilization is another intersex condition in which the fetus is exposed to the hormone progestin, which converts to androgen, and "virilizes" the body. This means that a female infant may be born with an enlarged clitoris, or complete penis, and/or a fusing of the labia. The child will have ovaries and a uterus, but in extreme cases she may not have a vagina or cervix. CAH occurs in karyotype 46 XX individuals, and is one of the most common intersex conditions. Like Progestin Induced Virilization, an anomaly of adrenal function causes androgen to "virilize" the fetus in utero. *Klinefelter syndrome* refers to individuals who inherit two X chromosomes (from their mother or father) and one Y chromosome from their father. A man with *Hypospadias* has his urethral meatus (the opening from which urine is excreted) on the underside of his penis, rather than at the tip.

There are many more conditions that appear under the intersex umbrella. Intersex conditions represent a very interesting mix of physical and social elements. It would be difficult to argue, for instance, that intersex conditions are entirely socially constructed, given that their definition is so tightly dependent upon physical matter such as chromosomes, hormones, and genitals. On the other hand, intersex conditions clearly require social definitions to exist. Chapter 3 provided a discussion of the social scientific construction of skeletons, gonads, hormones, and genes as signifiers of "sex." Here I want to highlight another common signifier of "sex": genitals. Far from there being one "mould" from which all penises, testes, clitorises, labia, and vaginas derive, human animals (like other animals) display a fantastic range of genital appearance. As we will see in this chapter, intersex conditions are most often first identified when an infant is born and medical staff want to proclaim "it's a girl!" or "it's a boy!" Most people do not realize that medical practitioners in Western countries actually refer to a "genital grid" to determine whether a newborn's clitoris or penis is "normal." In Figure 7.1 Anne Fausto-Sterling shows the grid used in North America.[49]

According to this grid, medical practitioners consider a clitoris between 0.2 and 0.7 centimeters acceptable, and a penis measuring between 2.5 and 4.5 centimeters acceptable. A clitoris longer than 0.7 centimeters, or a penis shorter than 2.5 centimeters raises alarm for doctors, and may well result in surgical intervention. Medical practitioners arrive at these quantitative definitions of "acceptable" genitals by measuring the clitoris and penis length of various samples of infants to

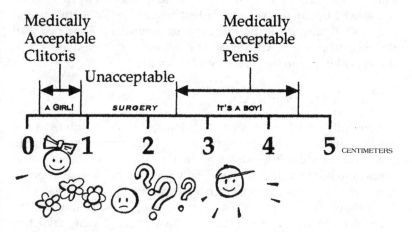

Figure 7.1 Illustration of "Phall-O-Metrics"

Source: Fausto-Sterling, A. (2000). *Sexing the Body: Gender Politics and the Construction of Sexuality.* New York: Basic Books, p. 59. Reprinted by permission of the Perseus Books Group.

determine the "average" length variance, and then define this "average" as "normal." Imagine if medical practitioners did the same thing with height. The "average" height of a female individual in North America might be 5 feet and 6 inches, and doctors supposed that there was a strong correlation between height at birth and eventual adult height. If doctors measured a female infant at birth and determined that she was "too short," would we routinely perform repeated operations on this infant in order to lengthen her legs? In fact, doctors do not usually perform surgery on people to increase their height because society *expects* height variation. Precisely because variations in genital size and appearance are not really thought about, discussed openly, or expected, most people do not realize that medical practitioners in Western countries routinely make judgments about the "appropriateness" of infants' genitals.

I am taking care here to refer to Western approaches to children with intersex conditions, because these approaches are distinctly Western. Asian, African, and other non-Western cultures approach intersex quite differently. Most markedly, non-Western cultures do not routinely medicalize intersex conditions, and so do not advise surgical and hormonal treatment. But in order to understand current Western approaches to intersex, we must understand how intersex itself was created. The following section details the emergence of intersex as a category, and eventually, a "problem" in need of solution. This section is followed by an overview of current instantiated Western medical protocols. Finally, I will consider the host of problems associated with this medical protocol, with a view to considering again sex diversity in human animals, and the issue of sex "differences".

A short history of intersex

The individual with an intersex condition has never constituted a free form absent from normative discipline.[50] Contemporary analyses posit that shifts in societal responses to intersex should be understood in terms of incremental increases in negative disciplining. Social historians note, for instance, that people with intersex conditions first began to be medicalized in the sixteenth century (Epstein, 1990; Foucault, 1980; Hausman, 1995). Ambroise Parel declared in 1573 that physicians should have the authority to determine the appropriate sex for the person with an intersex condition. His writing germinates the idea that cultural sex-appropriate behavior should carry equal weight with genital appearance in decisions about sex reassignment (Epstein, 1990:108).[51]

During the eighteenth century, modern medicine challenged the doctrine of humors. At the same time, as Chapter 3 details, the first drawings of sexualized human skeletons appeared, and from this point research into sex differences accelerated (Schiebinger, 1989). The division of human bodies into male and female suited "the economic needs of heterosexuality ... [lending] a naturalistic gloss to the institution of heterosexuality" (Butler, 1990: 112). The social conditions of people with intersex conditions were not immediately affected by the curiosity of medical science. In fact they remained much the same "until the availability of surgical and pharmacological interventions [that] could control or create a public sexual identity for them" (Epstein, 1990: 106).

But it is not the case that the intersex body was *free* before medicalization. Instead what we find is a major shift over several centuries in the juridical disciplining of bodies with intersex conditions. During the Middle Ages and Renaissance, the Courts recognized people with intersex conditions through the granting of legal standing. At issue was the person with an intersex condition's civil status rather than any concern about anatomical structures *per se*. The ritualized sexing of an infant at birth served (as it does to this day) a variety of functions, all of which can be regarded as mechanisms of the social organization of patrilineage: marriage, property, and inheritance rights.[52] The law assumed an unequivocal opposition between females and males, which explains why people with intersex conditions were subject to legal imputations of fraud. Such charges arose from the threat of *usurpation*. That is, gaining access to those privileges and powers to which they were not entitled, and indeed were meant to be denied (Epstein, 1990).[53] However, people with intersex conditions, upon entering adulthood, were given the legal option of deciding their own sex although a medical examination was necessary to confirm the legitimacy of that choice. Once categorized, the decision could not be reversed (Fausto-Sterling, 1993; Foucault, 1980).[54]

As the next section will detail, historians delineate three phases that occurred during the Classical Age, which necessitated a series of epistemological and ontological shifts in the understanding of sex complementarity. The first phase understood the person with an intersex condition as *two sexes in one body*. One of these sexes dominated and sex assignment was based on this *"natural"* domination. Phase two considered there to be *one true sex*, decipherable only by physicians. Gender assignment was based on the physician's expert declaration of the individual's *true* sex. In the most recent phase physicians and the psychiatric community conjoin *expertise* to uncover the *best sex* appropriate to morphology, psychology and, as I will argue, expediency. Through this medical and psychiatric configuration, the *material signifiers* of the body are altered to conform to a binary sex code.

Because social boundaries rely upon difference and hierarchy (Epstein, 1990), the person with an intersex condition threatens to disrupt and blur the boundaries of "everyday social relations," by the potential to "profit from their anatomical oddities" (Foucault, 1980: ix). Materially forbidden to exist today, people with intersex conditions have been erased historically by the enforced delineation of one "sex" from the other. That is, the intersexed body potentially threatens the stability of a naturalized sex order because people with intersex conditions are essentially *(a)non* in such a system. I will conclude the chapter by signaling some recent strategies that have been suggested for the conceptualization of the body in a way that opens up possibilities for the legitimization of an intersex subjectivity. I will argue that although these strategies appear to open spaces for a radical reinterpretation of sex, upon closer examination, these strategies may be dependent upon the very sex order they seek to disrupt.

Monsters and society

> Madness only exists in society. It does not exist outside of the forms of sensibility that isolate it, and the forms of repulsion that expel it or capture it. Thus one can say that from the Middle Ages up to the Renaissance, madness was present within the social horizon as an aesthetic and mundane fact; then in the seventeenth century ... madness underwent a period of silence, of exclusion. It lost the function of manifestation, of revelation, that it had had in the age of Shakespeare and Cervantes ... it becomes laughable, delusory. (Foucault, 1961: 9)

According to Michel Foucault, insanity is the creation of society. In other words, it is not the individual who is insane, but society that determines the classification of individuals as such. To yoke together madness and intersex would seem, at first glance, to misrepresent the latter. Worse, it would appear to reinforce the modern perception that politicized intersex groups wish to confound. However, I will suggest that such an approximation is useful in understanding the shifting societal responses to intersex.

The history of madness demonstrates that the classification of the variety of individuals we now consider "insane" is a relatively recent invention.[55] It follows that the ways in which society interacts with madness is historically dependent. Up until the end of the Middle Ages, the "insane" mingled relatively freely among the "sane." Although the subject of comedy and ridicule, society "just as often emphasized the tragic aspect of madness" (Barchilon in Foucault, 1965: v). This tragic aspect was closely bound up with the association of insanity with death. The mad person was seen to take the "absolute limit of death" and turn it inward "in a continuous irony" (Foucault, 1965: 16). The insane, in effect, guarded the secrets of death and the human spirit.

In a similar way, the person with an intersex condition may be seen as guarding the secrets of "sex." In ancient times anatomical differences between women and men were used to ground the social development of individuals based on "sex." The person with an intersex condition confounded such bifurcation. During this time, a person with an intersex condition was included in the loose classification of "human monster" (Braidotti, 1994; Daston and Park, 1998; Foucault, 1997). These beings were considered a double violation of the natural order: half-human and half-animal, they were the subject of particular trepidation as they "combined the impossible and forbidden" (Foucault, 1997: 51). However, there is evidence to suggest that those individuals who were explicitly identified as having an intersex condition were considered to be *special* in some way, able to sense and partake of the joys and secrets of both "sexes."[56] Cary Nederman and Jacqui True (1996) argue that in the twelfth century, people with intersex conditions were regarded as a discrete "third sex," based upon Galen's second-century theory of body temperatures: a continuum of different body temperatures determining a person's "sex." Galen's theory of body temperatures was similar to Hippocrates's humoral theory of body fluids and Ambroise Pare's theory of body heat. Each theory posited the hotter fluid and/or temperature to produce a male body while the colder fluid and/or temperature denoted the female body.

However, by the end of the Middle Ages and the emergence of the Renaissance we begin to see a shift in approach to the insane and the person with an intersex condition within this classification.[57] The mad were separated off, "confined in the centre of social relations within a set of institutions which segregate unreason from reason" (Turner, 1987: 64):

[The] humanist praise of folly thus inaugurates a long tradition that will seek to define, control, and ultimately confiscate the experience of madness. This tradition tries to make of madness an experience in which the human being is constantly confronted with his [sic] moral truth, revealing the rules proper to his nature. (Miller, 1993: 102)

Punishment was replaced by other forms of regulation (Turner, 1987). Medicine and psychiatry provided key sites for such regulation. From the Renaissance to modern society, this continued regulation has taken on a myriad of forms through classification, internment (in prisons and later mental institutions), and *cure*.[58] The increasing regulation within medicine and psychiatry also partially disqualified the individual as a legal subject (Foucault, 1997; Szasz, 1970; Zola, 1972).[59] The intersex "monster" became less powerful; a figure more to be pitied than respected. This provided a discourse with which to pastorally regulate the person with an intersex condition through medicine and psychiatry.

Rendering sex diversity harmless: the development of medicine and the reconstitution of monsters

By the late Classical period, medical science understood that sex determination was an enormously complex mechanism that resulted in a wide variation of "sexual types." This variation stretched far beyond the two mutually exclusive categories of female and male. In spite of this, the notion that the person with an intersex condition constituted a combination of two sexes in one body became supplanted by an understanding of a single *true* sex disguised within an ambiguous body. The invention of microscopy, along with increasingly sophisticated surgical techniques enabled an examination of gonadal tissue in live patients. Developments in anesthesia and medical hygiene reduced the risks involved with laparoscopy, and it was within this context that Klebs classified *true* intersex as the existence of both testicular and ovarian gonadal tissue (Hausman, 1995: 78). Thus, the "*pseudo*-intersexual" came into being. The *pseudo*-intersexual was defined as a person with determinate gonads of either one sex or the other, but ambiguous genitalia. For instance, the female *pseudo*-intersexual was characterized by having ovaries and "masculinized" genitals (Epstein, 1990; Hausman, 1995; Kessler, 1990). This terminology necessarily presupposed the existence of a single dominant sex *in* the body.

This epistemological shift (see also Chapter 3) affected the intersex body profoundly, for the delineation of a "true" sex always underneath the surface waiting to be unshackled from the trickeries of nature, foreclosed the option of choosing a "sex." The emergence of a *one* "sex" per body model created the role of expert for physicians via the medical

examination: the ritual employed to decipher a *true* "sex." Thus physicians positioned themselves as the *doubly inscribed*: both discoverer and
determiner (Kessler, 1990), entering into a mutually reinforcing relationship with the law to regulate the intersex body.

Medicine, psychiatry, and new forms of discipline

The mid-1800s witnessed the emergence of new ways of being as a result
of a number of factors including the birth of the modern nation, colonization, and industrialization. The emergent bureaucratic and admin-
istrative controls – those regulatory modes necessary to capitalist
economy – were unfolding as Freud was developing his theory of the
unconscious. The birth of psychoanalysis as a discipline, in unison with
an increasing medicalization of the body provided new regulatory techniques and mechanisms for the management of the modern subject.
The modern regulation of "sex" was only possible once "sex" as a concept had come into being. Those early attempts to discover the *true*
"sex" of the person with an intersex condition were part of a larger project to unravel the mysteries of the newly created subjectivity called
"sexuality," in an attempt to "identify, classify, and characterize the different types of perversions. [Such] investigations dealt with the problem
of sexual anomalies in [both] the individual and the race" (Foucault,
1980: xii).

The "discovery" of hormones at the turn of the twentieth century her-
alded the birth of a new strand of medical science: endocrinology (see
Chapter 3). The ability of the endocrinologist to "fix nature's mistakes"[60]
represented a "victory for the ideal of humanity, victories for normality"
(Hausman, 1995: 26). Endocrinology gained significant ground as a
cultural discourse in the first half of the twentieth century, enabling
endocrinologists to partake in the management of the intersex body.
Much of the medical literature of the day concerning intersex argued that
treatment needed to equate with what would later be referred to as the "sex
of assignment." Evident in this literature is a consolidation of the notion
of "psychosexual identity," later coined by John Money as "gender identity" (Hausman, 1995). But the physician's role would remain to some
extent circumscribed until genital surgical techniques reached a certain
level of sophistication. The rapid acceleration in the development of
medical and surgical techniques along with the emergence of new understandings of subjectivity prepared the ground for a gradual shift in discourses around "sex" and sexuality. This shift would have a profound
effect on the modern Western person with an intersex condition.

The modern regulation of intersex may be characterized by the epistemological shift from the notion of a "true" sex that determined one's sexed "destiny," to that of *"best"* sex. That is, the "sex" deemed most appropriate or advantageous given genital morphology and psychological and social environment. Such a shift was only possible because of the advances in medical science and the development of the "psy" disciplines which made the notion of a sex change operation conceptually possible.[61] Once this conceptual shift was made, medical technology enabled physicians to invoke a gender binary through the material production of sex (Hausman, 1995; Hird and Germon, 2001; see also Chapter 2). It must be noted however, that this shift was not all encompassing. While *best* sex may underpin the rationale behind contemporary medical procedures, physicians continue to imply to parents that they are able to discover the child's *true* sex.

Since the 1950s, protocols for the treatment of infants born with genitalia that does not conform to the normative measurement criteria outlined by Fausto-Sterling's "phallo-metrics" earlier in this chapter, have been developed from the work of John Money. Money's work is central to understanding the development of modern medical discourses on intersex. Indeed, the concept of gender identity has had a profound effect on the medical management of intersex. Borrowing the term from philology, Money defined "gender" as the "internal representations" of what it is to be a male or a female (identity), in unison with socially prescribed modes of conduct for that identity (role). The appropriation of this terminology was a result of Money's perceived need for:

> [a] terminology that would permit me to write about their [intersexual's] sexual and procreative lives as male or female, *despite the handicap* of having been born with a *birth defect of the sex organs* and *despite the relative success or failure* of attempted surgical *repair of the defect.* (1985: 280, my emphasis)

It was not long before the term entered medical discourse, particularly with regard to homosexual people and people with intersex conditions. While gender *role* rapidly came to represent the socially expressive component of the original definition,[62] Money remained committed to the idea that these two referents (role and identity) constituted a synthesis and should not be separated.[63] We might expect that the emphasis on gender identity as *socially* acquired, might lead Money to conclude that anatomy is not destiny, especially since he is studying children with *variable* genitalia who nevertheless identify as either girls or boys. But Money reclaims the importance of alignment between "sex" and "gender" by defining a "critical period" of parent–child interaction which cements an earlier in-utero period, where hormonal activation of the brain sets the direction of neural pathways in preparation for the reception of "post-natal social gender identity signals" (Raymond, 1994: 47). His theory of gender development was thus premised on the notion of the acquisition of a *core* gender identity which occurs during the first two years of life and is considered to be a critical developmental period for the "differentiation and establishment of gender identity and gender role" (Money and Tucker, 1976: 51). According to Money, core gender identity results from the child's interactions with parents; the child's perception of her/his genitals; as well as some mysterious kind of "biologic force" (1985: 282). This core identity builds upon itself as a result of cumulative experience. The critical period is said to cement an earlier in-utero period, where hormonal activation of the brain sets the direction of neural pathways in preparation for the reception of "post-natal social gender identity signals" (Raymond, 1994: 47). Using these concepts, Money was able to provide the rationale for the imperative to intervene as soon after birth as possible, *for the child's psycho-social well-being.* But Money's theory is highly contradictory. On one hand, his theory clearly privileges socialization, to the point where socialization becomes destiny. As a result, his work promotes the inscription of the social (a gender) onto the biological (intersex body). On the other hand, the primacy of the appearance of the genitalia reveals a quite crude recourse to the primacy of the biological body, as the following discussion suggests.

The arrival of a newborn with ambiguous genitalia is considered a "social emergency" (Pagon, 1987). Referring to Money's protocol, "sex" reassignment surgery is carried out as soon as possible after birth. Chromosome tests determine the genetic make-up of the child: if they reveal an XX configuration, genital surgery is usually performed without delay (Kessler, 1990). Chromosomes are somehow more meaningful for girls than they are for boys, as illustrated by the way that test results indicating the presence of two X chromosomes provide an immediate mandate for removal of the phallic/clitoral tissue. Implicit in the medical literature and the treatment protocols is a privileging of maleness, and an undervaluing of femaleness. Delays in "corrective" surgery to reduce (or remove) phallic/clitoral tissue of an XY infant beyond the neonatal period is to invite "traumatic memories of having been castrated" (Kessler, 1990: 8). Clitoroplasty, on the other hand, is undertaken when the child is anywhere between seven months and four years of age, and sometimes as late as adolescence.[64] Further, little attention is paid to aesthetics in the creation of a vagina. The sole (heteronormative) requirement is that the vagina be able to accommodate a penis.[65] Scar tissue is often hypersensitive resulting in extreme pain during intercourse. Because of the lack of elasticity in scar tissue, a daily regime of dilating the vagina is required to prevent the vagina from closing. The vagina is often constructed using bowel tissue, which lubricates in response to digestion rather than arousal (Laurent in Burke, 1996). An example of the level of expediency involved in surgical intervention is illustrated in the following quote by two key British surgeons:

> The clitoris may be reduced in size in various ways. The simplest and probably the most satisfactory is to remove it by amputation ... the dorsal [erotic] nerve of the clitoris runs within the sheath of the corpora cavernosa and it does not seem that its dissection and preservation are practicable. Whilst in theory preservation of the glans has something to commend it, the results of *amputation appear satisfactory*. (Dewhurst and Gordon, 1969: 41, my emphasis)

Whether genitals, hormones, or chromosomes are preferenced in "determining" an infant's "sex" is debated.[66] Given the salience of visual cues, the "abnormal" appearance of a newborn's genitals most often initiates medical intervention. In the first instance, then, genital appearance is privileged over hormones, chromosomes, gonads, and internal sexually reproductive structures (Hausman, 1995). Garfinkel and Stoller argue that the "natural, normally sexed person" as cultural object, must possess either a vagina *or* a penis and where nature "errs," human-made vaginas and penises must serve (1967:122–123). Surgical and hormonal treatment interventions are deployed in an attempt to ensure that the subject's body conforms to the assigned gender (Hausman, 1995; Raymond, 1994).

Where tests indicate the presence of a Y chromosome, surgery may be delayed while further tests determine the responsiveness of phallic tissue to androgen treatment. Such treatment serves to enlarge the penile structure to the point where it can pass as a *real* penis:

> Since ... reproduction may be disregarded, the most important single consideration is the child's subsequent [hetero] sexual life. ... If there is little or no penile growth the male sex will be out of the question and the female sex should be chosen; with good penile development the male sex may be appropriate. (Dewhurst and Gordon, 1969: 45)

The old trope proves true as penis size ultimately dictates whether the child is reconstructed as male or female (Griffin and Wilson, 1992; Pagon, 1987). Surgeons consider the condition of a micro-penis so detrimental to a male's morale that reassignment as female is *justified on this basis alone.* The implication here is that *male* "sex" is not only, or most importantly, defined by chromosomes or by the ability to produce sperm. Rather, masculinity is determined by the aesthetics of an *appropriately sized* penis:

> If the subject has an inadequate phallus, the individual should be reared as female, regardless of the results of diagnostic tests. In the patient with an adequate phallus, however, as much information as possible should be obtained before a decision is made. (Griffin and Wilson, 1992: 1536)

Consequently, it is common for infants with an XY chromosome configuration to be assigned and raised as female.

The high profile John/Joan case through which Money first argued for the necessity of surgical intervention illustrates many of the contradictions in the modern two-sex model of sex differences. After a bungled circumcision during infancy, John eventually found himself in the hands of Money's surgical team, who reassigned him as female. John's case was particularly important because he happened to have an identical twin brother. Money argued that if John "lived" the experience of "femaleness," then sociality, not chromosomes, determine "gender" identity. While Money repeatedly detailed the success of Joan's living as woman, interviews with John since he became an adult reveal this success to have been greatly exaggerated (Colapinto, 1997). Despite Money's assurances that *Joan* would live comfortably as a woman, *John* now lives with his wife, three adopted children and a reconstructed penis, adamant that he is a man. Tragically, John Reimer committed suicide in May, 2004. While John and Money would seem to disagree on just about every "fact" of this case, they concur as to the constitution of femininity and masculinity. Money argued that the identical twin brother was "male" because he preferred playing with "cars and gas pumps and tools" while John was "female" because of his preference for "dolls, a doll house and doll carriage." John himself says that he "knew" he was not a girl because, among other signs, he did not like to play with dolls, preferred standing while urinating, and daydreamed about being a "21-year-old male with a moustache and a sports car, surrounded by admiring females" (1997: 69). The various psychiatrists who eventually examined

John also used similar markers to define his "underlying" masculinity. One psychiatrist, for instance, described seeing John "sitting there in a skirt with her legs apart, one hand planted firmly on one knee. There was nothing feminine about her" (1997: 70). Paradoxically, at the same time that the medical community strongly requires a biological definition of an intersexual's "sex," the surgeons, endocrinologists and psychiatrists themselves clearly employ a *social* definition.[67]

Defining "Truth"

The modern medico-psychiatric response to intersex reflects a particular power–knowledge relationship. That is, "when scientists look to nature, they [ultimately] bring with them their sociopolitical beliefs about what is natural" (Spanier, 1991: 330). This produces a self-referential process: creating, reflecting, and reinscribing. This reinscription, or regulation, is an attempt to bring into discourse the very things that seem to escape discursivity, what poststructuralists have come to call "the real." Therefore, the codes created by scientific endeavor have a different relation to "the real" than other kinds of codes (Hausman, 1995: 24).

In other words, science has historically recognized the diversity inherent in "sex" diversity across many animal and plant species, including humans.[68] Despite this, modern discourses produce a specific knowledge about what is "*natural*" about "sex." That is, that "sex" consists of two mutually exclusive typologies: female and male. This sociopolitical belief "is maintained and perpetuated by the medical community in the face of overwhelming physical evidence that this taxonomy is not mandated by biology" (Hausman, 1995: 25). While physicians dealing with infants with intersex conditions are willing to partially accept the idea of plural "sex," they continue to subscribe to a binary notion of "sex" as *the* necessary code. Doctors "are willing to uphold binary gender by producing binary sex, ... [and use] technology to enforce binary gender by making males and females out of intersexuals" (1995: 77). It becomes clear then, that the *authenticity* of "sex" resides not on, nor in the body, but rather results from a particular nexus of power, knowledge, and truth. As Chapter 1 outlined, experts come to *define* the *truth* by virtue of having *knowledge*. Those experts then proceed to *discover* the *truth*, again, produced by a particular relation to *knowledge* (Foucault, 1980). In doing so they are doubly inscribed, as discoverer and as determiner. Indeed, these *experts* have been able to produce a discourse that:

> became powerful both as a justification for medical practices and as a generalised discourse available to the culture at large for identifying,

describing, and regulating social behaviors ... If you weren't born into a sex you can always become one through being a gender.[69] (Hausman, 1995: 107)

That something as "natural" as sex can be, or indeed needs to be, produced artificially is a paradox that appears to have escaped the medical fraternity (Kessler, 1990). If it is possible to *become* a "sex," then surely there can be no such thing as a natural "sex." By what means does sex resist exposure as the "constrained production" that it obviously is?[70] Medical explanations to families of infants with intersex conditions focus on the idea of a *continuum of sexual differentiation*, pointing to the bipotentiality of early *neutral gonads*, through the use of terms like "variation" rather than "abnormality" (Epstein, 1990): "knowing the proper terminology and understanding that genital ambiguity results from *normal developmental processes* helps to allay anxiety and provide a basis for understanding the type of evaluation necessary" (Pagon, 1987: 1020). Clearly there is a strong paradox here between a medical protocol ostensibly based upon "biological facts," and a medical protocol that clearly acknowledges that these "biological facts" consistently reveal *sex diversity*.

Moreover, the medical obsession with constructing *pseudo*-male and female bodies from intersex bodies is driven by a heteronormativity. If we are to understand that sex serves as a regulatory mechanism of heterosexuality, then by extension, it is clear that heterosexuality is itself a regulatory mechanism: of sexual reproduction (see Chapters 2 and 5). Paradoxically, medical *experts* will often sacrifice sexual reproduction in the interests of heterosexuality in the *management* of people with intersex conditions. By doing so, they indeed lend a naturalistic gloss to the normative institution of heterosexuality.

Resisting sex complementarity

Given the network of techniques brought to bear on the intersexed body through modern medical discourses, it is somewhat incredible that people with intersex conditions in the West have managed to create any sort of positive identity outside these discourses. A number of political intersex organizations, such as The Intersex Society of Aotearoa/ New Zealand (ISNZ) and the Intersexual Society of North America (ISNA), are increasingly articulating a counter-discourse. These organizations' primary political objective is the abolition of unnecessary genital surgery. Both organizations believe that since the "authenticity of gender seems only to reside in the proclamation of the expert, then the power to proclaim an alternative is equally available" (Kessler, 1990: 25).

Reports emerging from people with intersex conditions suggest a very high price is being paid by these individuals in the name of "normal" "sex" development. In response to the re-release of John Money's *Sex Errors of the Body* (1994), Cheryl Chase, founder of the ISNA writes:

> I have spoken to scores of intersexuals. Not one is grateful for cosmetic surgery of the genitals imposed during infancy. We know that it is, in fact, unjustified meddling. On the other hand, the risk of suicide in those of us who have been abused and shamed by non-consensual, mutilating plastic surgery of the genitals is very real. (*Hermaphrodites With Attitude*, 1994/1995: 9)[71]

While medical reports of surgical "sex" reassignment are either devoid of follow-up studies altogether, or focus on short-term follow-up only, the standard medical protocol has largely reiterated the necessity of surgery for stable "sex" identification. But as David remarks:

> Surgeons paint a picture of nearly any proposed surgery as routine, low risk, highly successful, and only mildly uncomfortable ... Where are they? I desperately want to hear such stories. I feel ashamed that I have healed so poorly when "all those others" have done so well. Why is it that I cannot find information on these well-adjusted adults? (*Hermaphrodites With Attitude*, 1996: 8)

Short-term studies are relevant only to the question of the immediate physical success of a procedure. Those studies that assess long-term surgery results refer almost uniquely to vaginoplasty. "Success" is quite limited – this means achieving regular, heterosexual acts, involving vaginal penetration. The "success" claim by surgeons is mitigated by emerging reports from people with intersex conditions of their own postsurgery experiences. For Jean, "if orgasms before the recession were a deep purple, now they are a pale, watery pink" (in Chase 1994/95: 7). Morgan Holmes similarly reports:

> When doctors assured my father that I would grow up to have 'normal sexual function,' they didn't mean that my amputated clitoris would be sensitive or that I would be able to experience orgasm (or any pleasure at all). They were guaranteeing him that I wouldn't grow up to confuse the normative conception of who (man) fucks whom

(woman). All the things my body might have grown to do, all the possibilities went down the hall with my amputated clitoris to the pathology department. The rest of me went to the recovery room and I'm still recovering. (HWA, 1994/95: 10)

Chase adds that if the doctors had suspected s/he would become lesbian "they might have lobotomized me" (p. 9).[72] IQ reports:

As a consequence of "reconstructive genital surgery" during infancy, I have no clitoral sensation, and have never been able to experience orgasm. After many years of denial, I had a severe emotional crisis, with suicidal feelings. (1995: 14)

In many cases, the intersexual body itself resists the socially and surgically enforced production of sex. This is clearly evidenced by the ongoing problems that surgeons face in their attempts at successful penile construction, and the (repetitively) routine complications of vaginoplasty, not to speak of the long-term effects of consuming huge doses of sex hormones. As these few narratives of people with intersex conditions offered here, and the increasing number available (Coventry, 2000; Dreger, 1999; Kessler, 1990; LeVay, 2000; Preves, 2003; Tapestry, 1999) attest, surgical reconstruction and the ongoing surgical and hormonal "treatments" often contradict the primary medical edict to "first do no harm."[73]

Not only does the intersexed community challenge society's ocular and heteronormative biases concerning sex identification (one must have a recognizable vagina *or* penis) but the assumption that stable "sex" identification is based upon genitals at all becomes suspect. Many individuals with intersex conditions identify as a stable "sex" without, or despite, surgery. Moreover, the fact that individuals who identify as intersex rather than female or male are perfectly capable of forming long-term "mature" intimate relationships questions the association between stable "sex" identification based on genitals as the foreground to "mature" relational pleasures. To the extent that individuals with intersex conditions do experience difficulties in forming and maintaining intimate relationships, it is clearly the result of negotiating society's response to sex complexity, which, in this regard, is similar to the ways in which lesbian, gay, and bisexual people must negotiate societal prejudice.

Given the pervasiveness of modern medical discourses, it is not surprising that political intersex organizations have resisted through a politics of inclusion. The freedom accorded to the person with an intersex condition in ancient times consisted of the legal right to choose which sex to *be*. Such a choice is reified in modern political movements. According to the ISNZ and the ISNA, because most genital surgery is *cosmetic*, it should be deferred until the individual concerned is able to make an informed decision, and provide informed consent. What is at stake here is "the right of each individual to make decisions about our own bodies, and to define ourselves" (Feinberg, 1996: 105). Given the extreme dependence of social, political, economic and cultural interactions in Western society on the categorization of "sex", it is hardly surprising that the ISNA argues that infants should be assigned a sex (either female or male) at birth.

Sandy Stone, offers a strategy whereby people with intersex conditions become conceptualized "as a *genre* – a set of embodied texts whose potential for *productive* disruption of structured sexualities and spectra of desire has yet to be explored" (1991: 296). Some people with intersex conditions are identifying as a third inter-"sex":

> I really have a place in the world. I really am a human being, a very valid human being. It's just wonderful. I am very proud to come out as an [intersex] person. The world has tried to make us feel like freaks. We have felt like freaks. I felt like a freak most of my life, but look at me. I'm just a human being just like everybody else. (Barbara in Preves, 2003: 133)

However, there is nothing in this textual interpretation that determines that intersex will challenge the available categories of "sex" because a "third sex" accepts the ontology of two sexes (female and male) in its definition. Another example of an attempt to consider gender possibilities with regard to intersex is that of Fausto-Sterling (1993), who argues for the expansion of gender categories in order to delineate wide-ranging variation within the "catch-all" term of intersex. She identifies three subgroupings, attributing to each the following terms: *herms* are those in possession of one testis and one ovary; *merms* comprise those with testes and some elements of female genitalia, but no ovaries; and finally, *ferms* who have ovaries and some degree of masculinised genitalia, but no testes. These subcategories are each deserving of consideration as a "sex" in their own right according to Fausto-Sterling. However as critics point out, such categorizing is something of a poison chalice given the uncritical acceptance of the concepts *true* and *pseudo* intersex,[74] which does little to seriously undermine the dominant order of sex "differences."

The intersex body itself may yet prove a site to explore the power-knowledge *truth* of "sex." As Hausman suggests, "we ... need to recognize the body as a system that asserts a certain resistance to (or constraint upon) the ideology system regulating it" (1995: 14). Hausman goes on to argue that, "in this sense we can read the body's resistance to 'gender' ... as suggesting that in a critical return to 'sex' we may find a way to destabilize 'gender' as a normal-ising narrative in the twentieth century" (1995: 200).

A *critical* return to "sex" does perhaps hold disruptive potential for the relation of the expert to the subject. If the transgression of the *uncorrected* person with an intersex condition brings any "truth" to society, it is in how our common sense understanding of the meaning of "sex" is constructed. The experiences of people with intersex conditions challenge both the medical community and any theory that relies upon a biological notion of "sex" as bifurcated, to the extent that both are predicated on the "sex"/ "gender" binary to operate (see Chapter 2). To effect the incorporation of an intersex body surgically assigned as "female" involves a determination as to the constitution of femaleness. Any definition of "woman" that retains any corporeality must be able to define that corporeality and this is exactly where the problem begins in definitions based on "sex" (Hird, 2000, 2002; Hird and Germon, 2001). A woman with an intersex condition will have any combination of partially or totally surgically created vagina, labia, and/or breasts. She may or may not be able to sexually reproduce. If being female does not entail the possession of particular anatomical parts, then the artificial creation of these body parts is inconsequential. But our current assumptions about the constitution of "sex" struggle with such a reality.[75] As we have seen in Chapter 3, one commonly assumed criterion of "womanhood" is the presence of "female" chromosomes. I do not agree that "we" know we are born with "female" chromosomes. Most people do not bother to have their chromosome configuration checked for authenticity, so there are likely to be many more individuals with "ambiguous" chromosome configurations than we currently identify.[76] The adult(s) present during our births took a cursory glance at our genitals and defined our "sex." The growing political intersexual community identifies many of these problems: the variability of "sex" identification, the *a priori* assumption of "feminine" and "masculine" behavior, the phallocentric bias in sex reassignment, and the problem people with intersex conditions often experience in "belonging" to sexually identified communities. In making these claims, the ISNA necessarily keys into the wider debate about the "nature" of "sex."

The "materialist" recourse to "the body" based upon chromosomes, reproductive function, genomal makeup, and genital appearance turns out to be, after all of our efforts, quite superficial. As Chapter 5 detailed, beneath the surface of our skin exists an entire world of networks of bacteria, microbes, molecules, and inorganic life. These networks take little account of "sexual difference" and indeed exist and reproduce without any recourse to what we think of as reproduction. I am not arguing that bodies are unimportant, immaterial, mere chimeras or that all is "surface" (indeed, the surface of the body is *least* important, biologically speaking). Bodies are important and certainly "material," but not necessarily in ways which justify continued emphasis on sex "differences." Rather than turning away from the "materiality" of the body, by looking below the surface of the body to its matter we find that intersex conditions are part of the natural "sex" diversity produced in any healthy living species:

> If a person of my condition defines themselves as neuter, you're basically defining yourself by what you are not, and then you're less than. I don't feel that I'm less than. I don't feel that I'm a genetic mistake. I don't feel that I'm genetic junk. I don't feel that I'm a genetic failure; [I'm a] genetic variation. (Meta in Preves, 2003: 127)

Suggested readings

Chase, C. (1998) "Affronting Reason," in D. Atkins (ed.) *Looking Queer*. New York: Harrington Park Press, pp. 205–20.

Dreger, A. (1998) *Hermaphrodites and the Medical Invention of Sex*. Cambridge, MA: Harvard University Press.

Fausto-Sterling, A. (2000) *Sexing the Body*. New York: Basic Books.

Hird, M. (2000) "Gender's Nature: Intersexuals, Transsexualism and the 'Sex'/'Gender' Binary," *Feminist Theory*, 1(3): 347–64.

Kessler, S. (1998) *Lessons From the Intersexed*. New Brunswick, NJ: Rutgers University Press.

Preves, S. (2000) "Negotiating the Constraints of Gender Binarism: Intersexuals' Challenge to Gender Categorization," *Current Sociology*, 48(3): 27–50.

The International Foundation for Gender Education (1999) Transgender Tapestry. Focus: Intersex, Issue 88. Waltham, Massachussetts.

8
How to Have Sex without Women or Men

> Some transformations are overt and heroic; others are quiet and uneventful in their unfolding, but no less significant in their outcome.
>
> (Gould, 2000: 79)

One long argument

In the first line of the last chapter, Charles Darwin described *The Origin of Species* as "one long argument" (1998: 346). The same can be said of the present book, that it is "one long argument" concerning the limits of "common sense" and scholarly assumptions about the "nature" of "sex." This book had two aims. The first aim was to outline the social study of science and nature in relation to "sex," sex "differences," and to a certain extent, "sexuality." I argued that our understandings of "sex" are based less upon biological knowledge of morphology and more on a sociocultural discourse that emphasizes sex dichotomy. In Chapter 2, I outlined the development of this sociocultural discourse as both an epistemic shift from knowledge through revelation (religion) to knowledge through systematic observation and induction and deduction (science), and a political shift toward sex complementarity through which a hierarchical order of privilege was maintained between women and men. I reviewed scholarly work that argues women and men were governed according to a "one-sex" model, in which femininity and masculinity appeared on the same axis. I argued that while women and men were not "free" from regulation, the "one-sex" model allowed a more fluid and less rigid understanding of "sex." The Enlightenment project undertook a fundamental revision of the meaning of "sex" such that femininity and masculinity eventually appeared as mutually exclusive entities. Through both epistemic and political shifts, a hegemonic discourse of "sex complementarity" figured women and men's morphology, intellects, emotions, and behaviors as opposite to each other and therefore complementary. For instance, the political development of the middle class notion of "the family" heavily utilized sex complementarity insofar as women became associated with the emotional and physical (childcare, cooking, and house cleaning) labor of the home and men became associated with the intellectual (decision making) and income generating (paid work) labor. Combining the proclivities of women and men together produced a complete, functional home. I detailed the shift from the "one-sex" to "two-sex" model in order to emphasize the point that our current understanding of sex "differences" was made possible through this significant epistemic and political shift.

The hegemony of science as the purveyor of knowledge/truth in modern secular society is suggested by "common sense" discourses that relate the "essence" of "sex" and sex "differences" to morphology. In Chapter 3, I examined the main sites of these purported morphological differences: gonads, hormones, chromosomes, and genes. The chapter explored the ways in which science and culture often work conterminously to reinscribe "sexual difference" on to the human body. The history of research reveals that an *a priori* paradigm of sex "differences" informed the ways in which skeletons, egg and sperm morphology and activity, and hormone and gene analyses prioritized particular research questions and interpreted the data to emphasize not only differences between female and male morphology (for instance in female and male skeletal structures) but also in activity (for instance, eggs as sluggish and passive; sperm as strong and mobile). Fausto-Sterling (1997) and other feminist scholars of science point out that scientists not only authorized the importation of cultural discourses of "sexual difference" in analyses of physical processes, but that these cultural discourses were imbued with conservative moral prescriptions about "appropriate gender behaviors" (1997: 57).

The first three chapters of the book not only recognize but elaborate the feminist and social scientific critique of "sex" and "sexual difference." Analyses that detail the social construction of scientific knowledge argue that scientific "facts" are socially mediated and can only be understood within their particular social and cultural milieu. Patricia Gowaty (1997b) outlines the "weak" version of this argument; that sociocultural mores introduce bias into scientific research which limits the degree to which scientific knowledge can claim objectivity. The "strong" version argues that insofar as science is based on a set of knowledge claims, it is necessarily limited by the parameters of this knowledge. The first three chapters of the book provide robust evidence that social discourses have certainly introduced bias into scientific research on "sex" in as much as these analyses have concentrated on establishing sex "differences" rather than sex diversity. As such, and with Gowaty, the book argues that " *'objective* knowledge' is an oxymoron" (1997b: 14). However, the book stops short of fully conceding the implications of the strong version of the argument, that it is both impossible and useless to utilize scientific knowledge in discussions of "sex."

In his analysis of the paradigmatic bias with which Charles Walcott classified the fossils of the Burgess Shale, Gould argues that Walcott erroneously concluded that he had discovered sexual dimorphism in *Opabinia*. Gould notes that "we observe according to preset categories, and often cannot 'see' what stares us in the face" (2000: 128). He further writes: " ... conceptual blinders can preclude observation, while more accurate generalities guarantee no proper resolution of specific anatomies, but can certainly guide perceptions along fruitful paths" (2000: 128). Despite his very thorough analysis of the bias of Walcott's classification scheme which dominated paleontology for decades, Gould does not eschew matter and science completely:

> We can argue about abstract ideas forever. We can posture and feint. We can 'prove' to the satisfaction of one generation, only to become the laughingstock of a later century (or, worse still, to be utterly forgotten). We may even validate an idea by grafting it permanently upon an object of nature – thus participating in the legitimate sense of a great human adventure called 'progress in scientific thought.' But the animals of the Burgess Shale are somehow even more satisfying in their adamantine factuality. We will argue forever about the meaning of life, but *Opabinia* either did or did not have five eyes – and we can know for certain one way or the other. (Gould, 2000: 52)

Along with recent feminist work that is actively pursuing knowledge about materiality, I agree with Gould that "the greatest impediment to scientific innovation is usually a conceptual lock, not a factual lock" (2000: 276); as such, the greater part of the book was devoted to making use of new perspectives in science studies. The second major aim of the book was to draw upon a loosely configured group of analyses termed "new materialism" to further contest cultural assumptions about "sex" and sex "differences." New materialism refers to a significant shift in the natural sciences that emphasizes openness and play within the living *and* nonliving world, contesting previous paradigms which posited a changeable culture against a stable and inert nature. I suggested these transformations within the natural sciences might be of interest to feminist social scientists who increasingly find themselves (often through "the body") grappling with issues such as "sex." On the whole, while feminism has cast light on social and cultural meanings of "sexual difference," there seems to be a hesitation to delve into the actual physical processes through which stasis, differentiation, and change take place. Recent studies suggest an enthusiasm on the part of feminist theory to revisit the issue of sex "differences," directly through biological science and throughout the book I sketch how some feminist scholars are working with matter to create science–literate analyses.

Chapter 4 outlined the major tenets of new materialism: nonlinearity and self-organization, contingency, variation, and diversity. Each of these principles is derived from Darwin's original theory of natural selection. I introduced a dualism that persists throughout the literature on evolution: conformity versus diversity. I argued that much of the development since Darwin's theory of evolution has emphasized law-like parameters on morphology and behavior dictated by nature, and I labeled these developments as the "conformity" side of the dualism. Neo-Darwinism and sociobiology tend toward conformity. In contrast, an emphasis on nonlinearity, self-organization, contingency, and variation means that new materialism tends toward the "diversity" side of the dualism. Stephen J. Gould argues that Darwin's original evolutionary theory is mainly concerned with the contingency of life (the "diversity" element of the dualism), emphasized by new materialism:

> Charles Darwin recognized this central distinction between *laws in the background* and *contingency in the details* ... The natural world is full of details, and these form the primary subject matter of biology. Many of these details are 'cruel' when measured, inappropriately, by human moral standards ... . And so, ultimately, the question of questions boils down to the placement of the boundary between predictability under invariant law and the multifarious possibilities of historical contingency. Traditionalists ... would place the boundary so low that all major patterns of life's history fall above the line into the realm of predictability ... . But I envision a boundary sitting so high that almost every interesting event of life's history falls into the realm of contingency. (2000: 290)

I argued that by drawing upon the principles of contingency, nonlinearity, self-organization, and diversity, feminists might analyze concepts such as "sex" and "sexual difference" that, while enjoying a history of social constructionist analysis, nevertheless persist in the sociocultural imagination as immutable and fixed. Kirby observes that contemporary critical analyses' insistence that the target of scrutiny is the discursive effects of objects, and not the object themselves, belies a construction of materiality as "rigid, prescriptive" and opposed to "cultural determinations that are assumed to be plastic, contestable, and able to invite intervention and reconstruction" (2001: 54). I argued this construction of the "rigidity" of materiality could be usefully challenged by the principles of new materialism; and by implication, the "plasticity" of cultural determinations might also be challenged. That is, I suggested that some of the most promising critical analyses of "sex" come from studies of materiality, and not the more well-worn cultural analyses with which most feminist scholars are perhaps more familiar. In short, I argued that new materialist studies of the matter of "sex" do not provide evidence for sex dimorphism, heterosexuality, or sex complementarity – since they are not hegemonic in nature, new materialism determines these practices as socioculturally determined. By implication, I hoped to challenge some feminist assumptions that the study of science and matter can serve only the interests of patriarchal attempts to maintain sex complementarity.

Chapter 5 focused on the nonlinear evolution of human "sex." It challenged the *a priori* acceptance within feminist theory of "sexual difference" based upon sexual reproduction. Drawing upon data from new materialism and nonhuman animal studies, Chapter 5 argued that the current recourse to "the body" based upon sexual reproductive function selectively attends to one aspect of "materiality" – that is, human bodies (like all other living organisms) engage in constant and varied reproduction, and only a small proportion is sexual. The chapter concluded by arguing that "nature" has been erroneously called upon to support the "truth" of sexual difference based on sexual reproduction. The aim of Chapter 6 was to introduce what is known as the "quiet revolution" (brought about by new materialism) in biology – that is, the diverse range of sex "differences," and sexual activities in strong species and ecosystems. The chapter reviewed how heteronormative assumptions about "sex," gender, and sexuality have influenced traditional biology to erase and silence sex diversity among living matter. This chapter argued that the vast majority of species display a diverse range of sexes and sexual activity, and documented lesbian and gay parenting, lesbianism, homosexuality, sex changing, and other behaviors in animals, plants, fungi, and bacteria. Finally, Chapter 7 applied the same principles developed in the previous chapter to the study of "sex" in human animals. Cultural notions of "sex" and "sexual difference" still maintain that when we take the human body as an autonomous entity, it is clearly sexually differentiated. People with intersex conditions provide a valuable opportunity to explore the relationship between "sex" and "gender," as well as the designation of meaningful categories of "sexual difference." Given the superabundance of sex diversity among animals, plants, fungi, and bacteria, the prevalence of intersex in human beings should not be particularly surprising. As such, it is particularly relevant to reflect upon the current silence surrounding Western society's attempts to eradicate intersex from the human population. This chapter sought to bring together analyses of intersex in its material (morphologic) sense with the ways in which intersex is "managed" within a cultural tradition that recognizes only two "sexes." The current "management" of intersex in Western culture reveals that the *authenticity* of sex resides not on, nor in the body, but rather results from a particular nexus of power, knowledge, and truth. People with intersex conditions' experiences of "sex" challenge Western society to the extent that it is predicated on the sex/gender binary to operate.

In sum, the preceding chapters argued that bodies are important and certainly "material," but not necessarily in ways which justify continued emphasis on "sexual difference." In the next section of this final chapter, I want to take a somewhat tongue-in-cheek examination of bacteria to illustrate both the evolutionary history of sex diversity, and its prevalence within the living world. As the oldest surviving living matter on this planet, and as the evolutionary origin of all living organisms, bacteria actualize "sex" in its ultimate diversity, defying cultural understandings of sexual dimorphism and sexual reproduction. I use the example of bacteria to emphasize the point that human cultural regulatory discourses surrounding "sexual difference" are particularly limited compared with the sex diversity evident in nature.

A bacterial ontology?

Biological studies suggest that our cultural reification of "sexual difference" is based upon a cursory and superficial understanding of organic materiality. Far from revealing sexual dimorphism, at every material level, our bodies practice a wonderful combination of intersex, reproduction and heterogeneous exchange with our environment. It is ironic that homogeneity in religion, nationalism, sexuality, race, ethnicity, and gender is so often encouraged over the heterogeneity we need to physically survive. I want to conclude this book by reflecting upon the human condition from a non-humanocentric perspective. By paying attention to nonlinear biology it is possible to acknowledge that human bodies, like all living matter, physically actualize sex diversity.

Taking account of our bodies as engaging in constant non-binary sex precipitates a reconsideration of matter, the integrity of the self, "sexual difference" and reproduction. On this point, I cannot resist ending with the observation that in our collective action (doing), human beings resemble beings that humans ironically revile – an argument made more pointedly by a computer (forced to live on earth among humans) than me:

> I'd like to share a revelation I've had during my time here. It came to me when I tried to classify your species. I realized you're not actually mammals. Every mammal on this planet instinctively develops a natural equilibrium with the surrounding environment. But you humans do not. You move to an area and multiply until every natural resource is consumed. The only way you can survive is to spread to another area. There is another organism on this planet that follows the same pattern. A virus. Human beings are a disease. A cancer of this planet. You're a plague, and we are the cure. (The Matrix, Warner Brothers, 1999)

Unlike Data from Star Trek, computer Agent Smith is to be reviled because it has the audacity to fear infection from humans. Perhaps Agent Smith's fear of infection stems from human beings' nearly unique propensity to "shorten all food chains in the web, eliminate most intermediaries and focus all biomass on themselves. Whenever an outside species tries to insert itself into one of these chains, to start the process of complexification again, it is ruthlessly expunged as a 'weed' " (De Landa, 1997b: 108). Indeed, humans seem to be the only species on this planet to fight against nonlinearity and diversity. In evolutionary and species survival terms, human beings most resemble viruses, which also survive by colonizing, and then consuming, new territories.

But rather than resemble viruses we might learn from another microscopic organism that displays a number of advantages over humans, especially with regard to reproduction. Fewer than 1 million days have passed since the birth of Christ (Margulis and Sagan, 1997: 14). Bacteria, on the other hand, have been around for about 3 billion years. Sagan is right to argue that "bacteria are biochemically and metabolically far more diverse than all plants and animals put together" (1992: 377). Bacteria are not half-hewn but fully living and evolved beings that have been thriving for more than 3500 million years. The greatest chemical inventors in the history of the Earth, they are not "just germs." Because of the conservative material nature of reproducing life, bacterial cells retain clues to the chemistry of Earth's surface as it existed in the remote past. Bacteria were the original hippies: they grew on nothing but sun, water, and air (Margulis and Sagan, 1995). Still the only beings able to perform many metabolic tricks of which we animals and even plants are not capable, bacteria were the first to breathe oxygen and to swim. They are the virtuosi of the biosphere. And they are also our relatives, which probably explains why we submit bacteria to such slander (Margulis and Sagan, 1995).

On their curriculum vitae, bacteria can boast that they are: the ancestor of all organisms on earth; the inventors of multicellularity (most bacteria are multicellular); the inventors of all major forms of metabolism; and the inventors of nanotechnology and metallurgy – 3000 million years ago, bacteria perfected the use of magnetite for internal compasses (Margulis and Sagan, 1995). But the list of bacterial accomplishments is far from complete. Bacteria can also detect light, produce alcohol, convert various gases and minerals, cross species barriers, perform hypersex, pass on pure genes through meiosis, shuffle genes, and successfully resist death. Bacteria are also the most important living beings to maintain the biosphere (Margulis and Sagan, 1995). Tabulating the vast number and diversity of bacterial accomplishments, Margulis and Sagan argue that human and nonhuman animal ability "has a long microbial fuse" (1997: 160).

Approaches that insist technology is the creation and purview of human animals are simply naïve: "we never invent anything that nature hasn't tried out millions of years earlier" (Clarke, 2000: 333). Life itself is, and has always been, "technological" in the very real sense that bacteria, protoctists, and animals incorporate external structural materials into their bodies (Margulis and Sagan, 1997). If, for instance, gene-splicing to create more socially desirable human beings is ever actualized, bacteria will have, by millions of years, beaten us to it by encouraging genes to cross species barriers. Current controversy over the use of animal cell and organ "donation" (no one, as far as I know, has ever asked the pigs for their consent) is old hat for bacteria. The equivalent to this bacterial ability in human animals would be a man with red hair and freckles waking up, after a swim with his brunette boyfriend and dog, with brown hair, a tail and floppy ears (Margulis and Sagan, 1997: 53). Much of human engineering, whether industrial or genetic, is borrowed, not invented: bacteria long ago cornered the market on "trans," whether

transduction or transfection. Much of the "brave new world" of reproductive technologies is human mimicry of well-worn, millions of year
old bacterial practices.

Our remote ancestors continue to promiscuously exchange genes
without getting hung up on sexual reproduction. Bacteria are not picky,
and will avidly exchange genes with just about any living organism
anywhere in the world, including the human body. Thus bacteria are
beyond the false male/female dichotomy of human discourse (Margulis
and Sagan, 1997: 89). Since bacteria recognize and avidly embrace
diversity, they do not discriminate on the basis of "sex" differences at
all. The bacteria that move freely into and within our bodies are already
infinitely "sex" diverse, as are most of the species on this planet. Because
of their extreme adaptability, which is enabled by their preference for
sex diversity, in evolutionary terms the most likely "species" to survive
on earth is indisputably bacteria. So in the tired game of identity, I
would choose neither goddess nor cyborg (Haraway, 1991). I would
rather be a bacterium.[77]

How to have sex without women or men

The title of this section comes from C. Jacob Hale's article "Leatherdyke
Boys and Their Daddies: How to Have Sex Without Women or Men"
(1997). This intelligent article argues that since "sex," "gender," and
"sexuality" are defined by sociocultural discourses, we can engage in
sexual relations without genders, or at least genders as they are norma-
tively defined. I have co-opted this title because it can also refer, through
the theoretical lens of new materialism used throughout this book, to
the argument that "sex" is not dichotomous. It makes as much sense,
biologically speaking, to talk about zero sexes (we are much more similar than we are different) or a thousand tiny sexes (to acknowledge the
symbiotic relationships bodies share with other bodies – bacterial or
otherwise, as well as the myriad of ways in which we reproduce other
than sexually) as it does to talk of two sexes. Moreover, in the symbiotic
relationships that literally sustain our lives, we are all "having sex" with
lots of other species. Plants and fungi have sex without females and
males. Many species of fish practice transsex. Plants and bacteria
routinely have transspecies sex. That culture focuses on two sexes is,
biologically speaking, arbitrary and it is ironic that biology is routinely
used in contemporary society to sustain the cultural notion of sex
dimorphism. But this focus on sex dimorphism is far from arbitrary
from an epistemological and political perspective: sociocultural

structures largely depend upon the discourse of sex complementarity. In the book, I used the dualism of "conformity" versus "diversity" to make the point that while nature emphasizes diversity, culture emphasizes dichotomy.

Because biology is routinely called upon to reify sex dimorphism, I suspect that many feminist scholars have eschewed the natural sciences as a useful site for critiques of this dichotomy. I have argued throughout the book that this reluctance to explore the natural sciences, and particularly biology, is counterproductive, since new materialism argues so strongly for concepts (contingency, nonlinearity, self-organization, and diversity) that are keenly supported by social constructionist (and particularly poststructural) analyses. My intention has been to provide a flavor of the possible intra-action between feminist theory and science studies. Of course, the arguments offered in this book support social constructionist arguments that seek to undermine the sex/gender binary insofar as new materialism offers ways to critique the "sex" part of the sex/gender binary. Rather than implicitly confirming that "sex" girders cultural notions of gender, my strategy has been to critique the girding itself. That is, by challenging "sex," we challenge not only assumptions about "gender" but the binary itself.

Suggested readings

Grosz, E. (1999) "Thinking the New: Of Futures Yet Unthought," in E. Grosz (ed.) *Becomings. Explorations in Time, Memory, and Futures*. Ithaca, NY and London: Cornell University Press.

Margulis, L. and Sagan, D. (1995) *What is Life?* Berkeley, CA: University of California Press.

Glossary of Terms

Autopoesis: From the Greek word "self-making," this term refers to self-maintenance by organisms. This ability is a prerequisite for reproduction.

Chimerism: Refers to the presence of two genetically distinct cell lines (genomes) in an organism. This may occur through inheritance, transplantation, or transfusion. See also definition in note 35.

Chromosome: Refers to a DNA structure (gene) made of chromatin (has a coil-shaped appearance). Most human cells contain 46 chromosomes.

Diploid: From the word "diploidy" referring to the state in which there are two sets of chromosomes in the nuclei of eukaryotic cells (i.e. two nuclei). Eukaryotic cells are cells that have a membrane-bounded nucleus.

Gamete: Refers to a haploid reproductive cell whose nucleus fuses with another gamete during fertilization.

Gene: Refers to a strand of DNA found on a particular location on a specific chromosome.

Haploid: From the word "haploidy" referring to the state in which there is one set of chromosomes in the nuclei of eukaryotic cells (i.e. one nucleus).

Heteronormativity: Refers to the hegemonic discursive and nondiscursive normative idealization of heterosexuality. Heteronormativity both established and maintains sex complementarity.

Hormone: Refers to a chemical compound released by one cell that travels through the circulatory system to affect the activity of other cells in the organism.

Meiosis: The process that occurs during fertilization whereby diploid cells are reduced to haploid cells (i.e. two nuclei are reduced to one nucleus, as in sperm and egg cells).

Mitosis: The process of doubling chromosomes. This form of cell division accounts for the reproduction of protoctists and the growth of animals, plants, and fungi.

Mosaicism: Refers to patches of tissue on an organism that differ genetically.

Natural selection: Refers to the central tenet of Darwinian evolutionary theory. Refers to the preservation of favorable variations and the rejection of injurious variations over a very long timescale.

Organic chauvinism: Manuel De Landa's term for the tendency of human beings to mistake the "purpose" of life to be their own organic being. De Landa argues that bodies are only temporary assemblages of the processes of life.

Parthenogenesis: The development of an organism from an egg in the absence of fertilization (by two parents).

Sex complementarity: Refers to the emphasis on differences between females and males rather than similarities. These supposed differences are structured to complement each other to form a functional whole (as in, for instance, the heterosexual family unit).

Sex/gender binary: Refers to a concept used with regularity in the social sciences whereby "gender" refers to socially constructed identities based upon biological "sex" differences between females and males.

Notes

Chapter 1: Introduction

1. I use the term "materialism" to refer to living and nonliving matter, rather than the perhaps more familiar definition of materialism as the social and economic relations between women and men. See Sheridan (2002) for a useful summary of the debate within feminism between the latter definition of materialism and cultural analyses.
2. My experience is not unique. Carla Golden writes: "I find that students are very open to the idea of gender as socially constructed ... but resistance to the idea of sex as social construction remains strong" (2000: 31).
3. One trajectory of the female epistemology argument suggests that the scientific approach to nature is itself fundamentally masculinist, and is one of the primary techniques deployed in the material oppression of women. For some (Bleier, 1984; Harding, 1986, 1991) this requires the fostering of distinctly feminine approach to materiality. For instance, one of the strongest proponents of a feminine epistemology argues the case through the materiality of fluids. Irigaray (1985) argues that masculinist science tends to favor solid objects in its descriptions of the world because solid objects give the appearance of "things in themselves." Fluids (and presumably gases) deny the fixity of solid objects, and introduce "things" as processes into the field of nature. As Olkowski describes, "if solids fail to adequately account for fluidity as a physical reality, then there must be something physically real, and that 'thing' must be fluidity, which ... is not a process but a thing" (2000: 78). Moreover, because fluids deny the characterization of substances or universals, they demonstrate a logic outside of the universalization of masculinist logic. For Irigaray, the potential of this alternate logic, nonlogic as it were, is the potential of a feminine epistemology. Others (Stengers, 1997, 2000) reject any notion of a "feminine science" and criticize the current feminist focus on the culture of matter as the predominant intervention into scientific enquiry (Wilson, 1996, 2000).
4. Some feminists would go so far as to locate certain "facts" outside of cultural construction. For instance, Marcy Lawton, William Garstka, and J. Craig Hanks state, "there are facts in science that are outside the bounds of narrative; the earth revolves around the sun (and not the other way around), regardless of what a particular text or even culture teaches, and this fact cannot be deconstructed" (1997: 65).
5. Bruno Latour refers to this as comprehending the "thingness of the thing" (2000b: 112).
6. I am not arguing here that biology and sociology have never been connected, or that all sociologists and feminist theorists lack knowledge about the natural sciences. My aim is to encourage greater matter "literacy" on the part of social scientists generally, as well as a more sustained critique of the assumption that the social sciences are limited to cultural analyses.

7. In some sense, debates about technology center around the separation of culture from nature. Jon Ward explains by recounting a conversation he had with a friend about homosexuality. His friend warned, "It's all very well, you know, but you can't argue with biology." Ward asks, "Has he never fried an egg? The whole of human history is an 'argument with biology' " (1987: 162–3). Frying an egg is a manipulation of "nature" through technology, and yet this taken-for-granted act is not considered "unnatural" in the way that homosexuality often is.

8. The reason that social scientists tend to side so ferociously with history as against nature is because in history social scientists see hope – things have not always been this way, which means that things do not have to be this way now or in the future. Nature, by contrast, signifies immutability, and for social scientists, this tends to be interpreted as all that is conservative. Moreover, Londa Schiebinger points out that rather than seeing science as *inherently* masculine, modern science became masculine as a result of the political separation of social and intellectual labor between women and men during the Enlightenment (1993: 233).

Chapter 2: Making Sex, Making Sexual Difference

9. I thank Jack Veugelers for redirecting my attention to de Beauvoir's early work on biology.

10. This was also true for transvestites. A woman's desire to approximate the male through dress was considered "healthy" and normal compared to the abomination against the natural order of male supremacy that the male desire to don women's clothing intimated (Bullough, 1974).

11. Judicial hearings detail behaviors such as dress, posture, language, and role assumed during intercourse as important indicators of the individual's "gender."

12. Laqueur (1990) shares a poignant and reflexive story of watching his father, a pathologist in the 1930s, painstakingly examine bodily organs taken from autopsies. To demonstrate that the body does not express its "truth" to be "read," Laqueur admits finding in his father's study a paper entitled *Further Studies of the Influence of Various Hormones on the Masculine Uterus.*

13. Lest we think this an outdated sentiment, note the statement by Aleen Quist, Republican candidate for Minnesota governor in 1994: "You have a political arrangement, and when push comes to shove, the higher level of political authority … should be in the hands of the husband. There's a genetic predisposition" (Los Angeles Times, July 12: 7).

14. For a useful summary of feminist critiques of the "sex"/"gender" binary see Sedgwick (1991), especially pp. 27–35.

15. Sedgwick (1995) makes the point that the rhetorical use of binaries does not only designate two discrete opposites, but also organizes the multiple differences between the two axes of any binary.

16. Shildrick seems to sidestep the question of sexual difference altogether as she lengthily quotes Haraway's (1991) work on cyborgs and then concludes "whether a cyborg has a sex is perhaps rather more complicated" and ends the article by stating "there remains also the issue … of what becomes of sexual difference" (1996:10, 12).

17. The Michigan Womyn's Music Festival has continued to exclude transgendered women from the physical space of the festival since 1991. The current policy is to include womyn-born-womyn-only, but the Festival organizers have dropped the *physical examination* required for admission to earlier Festivals (http://www.camptrans.com). Witness also the debate generated within the feminist community when the British Columbian Human Rights Commission recommended that "gender identity" be included as a discriminatory grounds for exclusion. The issue for some feminists was the inclusion (not surprisingly) of transsexual women, who were not understood to be "real women" in women-only spaces such as Rape Crisis Centers. Little discussion took place about transsexual men, who by this line of reasoning, should be allowed into women-only spaces because they are "real women."

Chapter 3: The Body of Sexual Difference

18. Interestingly, the same procedure of combining female and male forms is being used today in the Visible Human Project.
19. Rothblatt (1995) points out that the average Japanese man and American woman share the same weight and height, although we do not organize international sports tournaments such that Japanese men compete against American women, or that Asian men compete as a separate category.
20. Annemarie Jonson (1999) reviews arguments supporting the opposite claim that proteins produce DNA.
21. Mitochondria are theorized to contain DNA because mitochondria used to be independent living bacteria that millions of years ago were incorporated into other living organisms (see Margulis and Sagan, 1997).
22. Testosterone is also released in other chromosome variations.

Chapter 4: New Materialism, Nonlinear Biology, and the Superabundance of Diversity

23. Gould (2003) makes the interesting point that the more recent focus on genes (see, for instance, Dawkins, 1989) as the ultimate locus of selection is really an extension of Darwin's principle of agency, rather than a radical revision of the theory of natural selection.
24. Gould makes the important observation that while the branches of the "tree" of evolution do not join, hybridization between distant lineages does take place regularly in plants (making the tree resemble a complex assemblage more), and that genes transfer across species through viruses (2000).
25. Although Darwin is credited with both dramatically and fundamentally challenging the Christian structure of society by arguing against a divine origin of all living organisms and indeed the planet itself, Ruth Hubbard (1979) makes the point that in actuality much of Darwin's thesis not only accommodated the social thought of his times, but corroborated key ideas. Irvine argues that bourgeois ideals of the time consisting of "economic conceptions of utility, pressure of population, marginal fertility, barriers in restraint of trade, the division of labor, progress and adjustment by competition, and the spread of technological improvements" are all found in *The Origin of Species*

(in Hubbard, 1979: 13). Likewise, Hubbard argues that Darwin's theory of sexual selection is similarly biased toward the cultural mores of his day insofar as females are generally conceived as both passive and secondary to the males of any given species who take center stage.

26. For Jerry Flieger "Deleuze's great originality resides in his inmixing of planes or phyla – animal with plant and with human – and his explanation of 'transgression' in molecular terms, which apply to non-organic phenomena as well as to human and animal life" (2000: 44).

27. Take something that, at first glance, may seem simple: population density. Darwin stated that all animals have the capacity to reproduce more often than is needed simply to replace themselves. But most of the time, populations are held in check by a number of environmental factors such as predators, the scarcity of resources, and so on. We might predict that population density increases and decreases in a positive feedback loop. However, scientists have recently found that populations show an extraordinary variety of dynamical behavior. And some of this behavior is so random that it is chaotic (May 1989).

28. To illustrate this point, we typically assume that urban centers are suffering from an ever decreasing amount of "nature." Hence, the familiar routine of leaving the city's "rat race" on the weekend to search for "nature." But Nigel Clark argues that "wild" (organic) and "urban" (nonorganic) are far from exclusive categories. Not only do rats exceed human populations in any given city (literalizing the above slogan more than people generally assume), cities constitute dynamic systems of organic and nonorganic elements which vigorously combine to produce emergent properties. Mike Davis notes that various flora, fauna, and animal species including rats, coyotes, and raccoons all display unexpected and often chaotic resurrection within urban centers. In California, for instance, where the gourmet fed cat or dog, and the occasional jogger, has been prey to mountain lions, these lions appear to be in the process of "a behavioral quantum jump: the emergence of nonlinear lions with a lusty appetite for slow, soft animals in spandex (1998: 249)." See Clark (2000), Sprin (1984), and Davis (1998).

29. "Variations are not random in the literal sense of equally likely in all directions; elephants have no genetic variation for wings. But the sense that 'random' means to convey is crucial: nothing about genetics predisposes organisms to vary in adaptive directions. If the environment changes to favor smaller organisms, genetic mutation does not begin to produce biased variation toward diminished size. In other words, variation itself supplies no directional component. Natural selection is the cause of evolutionary change; organic variation is raw material only" (Gould, 2000: 228).

30. But the myth that sex dimorphism produces greater biodiversity is popular enough to have acquired its own nickname – *The Red Queen Hypothesis*. This name is derived from the Red Queen in *Alice in Wonderland* who tells Alice that she must run very fast in Wonderland just to stay in the same place. See Margulis and Sagan (1997).

31. For instance, when biologists theorize that a particular behavior is an adaptive trait in one species, there needs to be some explanation as to why that same trait is absent from a closely related species. Rowell uses the "dominance–subordinacy" relationship as an example. As much as hierarchical

relationships may stabilize group behavior through the increased predictability of individual behaviors, it would seem that social hierarchies are advantageous to group living (which all primates engage in). On the other hand, fighting is very conspicuous and involves the expenditure of a lot of energy, both counterproductive to adaptive living. Moreover, there is increasing evidence to show that high-ranking males do not have greater access to females for reproduction, nor do they necessarily produce more offspring.

32. A Derridean analysis might suggest that the culture–nature dichotomy is a precondition to the functioning of social constructionist arguments. Soper sees the debate as not arguing whether the distinction exists or not, but whether it is a distinction of kind or degree (1995: 41). I like the double-meaning of Frank Capra's remark "we never speak about nature without at the same time speaking about ourselves" (1975: 77).

33. In *Neural Geographies* (1998) Wilson meets the challenges of science literacy and analyzing matter as more than cultural representation. Here Wilson explores connectionism and cognitive theory. As a form of new materialism, connectionism stresses cognition as the connections *between* neuron-like units rather than cognition as "the manipulation of symbols in accordance with pre-existing computational rules" (1998: 6).

Chapter 5: The Nonlinear Evolution of Human Sex

34. The use of taxonomies has been critiqued by both social scientists (Foucault, 1994b) and natural scientists (Gould, 2000).

35. For an excellent feminist critique of the panic surrounding the apparent increase in estrogens in the environment see Roberts (2003a). For instance, Roberts demonstrates how much of the publicity and panic surrounding "sex" hormones is really about "female" sex hormones. Janet Raloff suggests that "with the growing ubiquity of pesticides and other pollutants possessing the functional attributes of female hormones, our environment effectively *bathes us in a sea of estrogens*" (1994: 56, my emphasis). Although Raloff writes about the environment, her concern with environmentalism is curiously absent when she questions, "What is the economic cost of having a generation that cannot reproduce?" (1994: 58). Roberts notes that, paradoxically, the literature tends to be much more interested in understanding female sex as the outcome of "lack" of "male" hormones than the possible effects of the mother's estrogens (erroneously termed "female" sex hormones) on her male infant.

36. Minerals and animals do not belong to separate kingdoms. All of the five kingdoms have species which produce minerals (see Margulis and Sagan, 1995).

37. A more extensive definition is provided by Chu *et al.*: "A mosaic is an individual with cell populations of more than one genotype (e.g. karyotype) derived from a single zygotic genotype through mutational or zygotic events (e.g. somatic mutation, somatic crossing over, mitotic nondisjunction, etc.). A chimera is an individual with cell populations of more than one genotype arising through a mixture of different zygotic genotypes (e.g. transplantation, chorionic vascular anastomoses, double fertilization, and subsequent participation of both fertilized meiotic products into one developing embryo, etc.)" (in Benirschke, 1981: 433–4).

38. For instance, contemporary social constructionist theory owes much to George Mead's theory of symbolic interactionism. Mead (1934) distinguished humans from all other animals through our supposedly unique ability to recognize ourselves as objects. Yet, recent studies conclude that chimpanzees and orang-utans recognize themselves, and subordinate simians hide copulation from dominant males (Margulis and Sagan, 1997). Language is another trait that human animals favor in distinguishing themselves as entirely unique and (usually) superior (see Chapter 6). However, all nonhuman animals communicate – indeed, the recent discovery of symbolic communication by honeybees "upsets the very foundation of behavior, and biology in general" (Griffin in Margulis and Sagan, 1995: 150). The homogenization of nonhuman animals is an attempt to shift attention away from the fact that humans share 98 percent of the same genes with chimpanzees. We must ask to what effect does such a taxonomy work? As Sarah Franklin notes "trading organismic distinction for pan-species genetic information flow pulls the rug out from under the sex/gender system as we know it" (1995: 69).

39. As Brown also observes, "how can we be so morally different if we're so physically similar" (1999a: 191)?

40. Recall Donna Haraway's example of "the one and the many" *Mixotricha paradoxa*, that engenders questions about the human notion of identity autonomy. See also Rackham (2000).

Chapter 6: Sex Diversity in Nonhuman Animals

41. Gowaty's points argue against the supposition that males will always choose young, nubile females. According to the principles of sexual selection, males will choose females who are most capable of looking after young successfully; that is, who will be "good providers, wily and competent at exploiting environmental resources on behalf of their offspring" (1997c: 99).

42. Interestingly, Kleinman (1977:40) notes that monogamy is among "the more highly evolved forms of social organization" but then also concedes that monogamy is much more common in birds (90 percent) than mammals (less than 3 percent).

43. The search for an association between hormones and homosexual behavior is predicated on the assumption that homosexuality is atypical and therefore the result of developmental processes "natural" to the "opposite" sex.

44. Dagg's compendium should be read as a conservative account because it only defines homosexual behavior if it is explicitly defined as such in the research studies it reviews.

45. Bonobos are an extremely interesting example of sexual diversity. Bonobos have been widely documented as engaging in a diverse range of individual, "same-sex" and "opposite-sex" behaviors. Bonobos also share a similar rate of reproduction with humans, suggesting that Bonobos have also separated sexual reproduction from sexual behavior (for more on Bonobo sexual practices see deWaal, 1995).

46. Dagg also makes a number of problematic statements about human behavior. She notes, for instance, that mounting of Buffalo during a stampede or

mounting by a rabbit during a fight "have no counterparts in human beings" (1984: 179) when this may clearly not be the case. She also states that homosexuality in humans "is frequent among the young, who later marry and become heterosexuals" (1984: 179). Given the complexity of human sexuality, it is problematic to assume that someone who is married is automatically heterosexual, or that the categories of "homosexual" and "heterosexual" are mutually exclusive.

Chapter 7: Sex Diversity in Human Animals

47. I defer to Kessler's (1998) distinction between genital "variability" and "ambiguity." See pp. 8–9.
48. For a fuller explanation of intersex conditions see www.isna.org/faq.html. Following Meyer-Bahlburg (1994), I use the term *intersex* to refer to both hormonal and nonhormonal categories of sex "ambiguity".
49. Different countries have different measurements for "acceptable" clitoris and penis length. (See, e.g. Lee *et al.* (1980); Oberfield *et al.*, 1989; Yokoya, Kato, and Suwa, 1983.)
50. For example, the ancient Jewish books of law (the Talmud and the Tosefta), contain extensive lists of regulations directed at people with intersex conditions. These include sanctions against inheriting property; shaving; serving as witnesses in legal trials; and entering the priesthood, among others (Fausto-Sterling, 1993: 23).
51. I use the term "reassignment" here because "nature" has already assigned the individual as having an intersex condition.
52. These included the prevention of fraud; the regulation and maintenance of differential privilege between males and females; and the regulation of morality and reproductive family life (Laqueur, 1990).
53. Voting is a powerful example of the mechanisms of heteronormativity, all the more so because of its apparent banality. Voting was the prerogative of (white) men – as citizens. Women and *others* were not citizens and were thereby prohibited from voting. Were persons of indeterminate sex allowed to marry, the sanctity of heterosexual marriage was threatened because such bonds carry the specter of homosexuality. If one partner's sex is ambiguous, then "the sex of both partners may be the same or ambiguous and therefore potentially the same" (Epstein, 1990: 129).
54. Harsh legal penalties were metered out to those caught transgressing. Legal injunctions targeted the outward manifestations of sex as a means of regulating the social behavior of people with intersex conditions (Butler, 1993; Epstein, 1990; Foucault, 1980; Kessler, 1990).
55. In ancient times, the "insane" served as an umbrella under which all sorts of individuals, including transients and people with intersex conditions, stood.
56. This is not to make the claim that this is how people with intersex conditions experienced themselves, only to posit reasons for their relative autonomy in society at that time.

57. The age of confinement, and the exclusion of the unreasoned corresponded with a philosophical shift in the work of Descartes whereby madness (unreason) was excluded from the domain of philosophy (Brown, 1985).

58. One of the most interesting and revealing developments in the psychiatric classification system was the classification of homosexuality as a mental illness, and then its subsequent liberation from this classification.

59. Szasz (1970) argues that the concept of insanity has provided the judicial system with a useful device for dealing with criminal offenders, providing the basis for stripping the subject of legal status. Furthermore, the term insane was (and is still) used to justify nonconsensual medical intervention.

60. Endocrinology is concerned with disease occurring *within* the human organism: produced *by* the body rather than by external contagions.

61. Nikolas Rose (1996) coined the term "psy" to describe the regulating disciplines of psychiatry, psychology, and psychoanalysis.

62. The term "gender identity" was adopted by clinical psychology to represent "a psychodynamic state of being" (Money, 1985: 282).

63. This is made explicit in the provocatively entitled article, "The Conceptual Neutering of Gender and the Criminalisation of Sex" (1985).

64. Despite increased risks of stenosis or injury that accompany early vaginal construction, some physicians "prefer" to complete all surgical procedures before the child reaches 18 months of age (Perlmutter and Reitelman, 1992).

65. Indeed, surgical teams consider that one of the worst mistakes that can be made is to "create an individual unable to engage in genital [i.e. heterosexual] sex" (Kessler, 1990: 20).

66. It is further testament to the variability of "sex" that several factors can be used singly or in tandem to "determine" an individual's "sex": chromosomal sex, hormonal sex, gonadal sex, genital sex, and sexually reproductive sex.

67. Kessler quotes one interviewed endocrinologist as saying "why do we do all these tests if in the end we're going to make the decision simply on the basis of the appearance of the genitalia?" (1990: 13).

68. In spite of the multitude of physical and physiological problems that arise from the (re)construction of intersexual flesh, "scientific dogma persists with the assumption that intersexuals are doomed to a life of misery without medical intervention" (Fausto-Sterling, 1993: 23).

69. However, evident in the emergent counterdiscourse of medically mediated people with intersex conditions is that the *source* of the trauma is not the experience of inhabiting an intersexed body but rather, from the experience of medical interventions. (See Bodeker, 1997, Burke, 1996; Holmes, 1995; Triea, 1996).

70. In *History of Sexuality* (1979) Foucault inverts the traditional understanding of the relationship between sexuality and sex. "Sex" has been understood as the root cause of the structure and meaning of desire (including sexuality). For Foucault, the body does not respond to some form of essential sex, creating desires, pleasures, and sexuality. It is sexuality, invested by power relations that "produces" sex.

71. *Hermaphrodites With Attitude* is a newsletter published by the ISNA. See www.isna.org.

72. Indeed, surgical teams consider that one of the worst mistakes they can be made is to "create an individual unable to engage in genital [i.e. heterosexual] sex" (Kessler, 1990: 20).

73. To my mind, the most succinct response of a person with an intersex condi-
    tion to forced gender reassignment surgery reports:

    I was most fortunate, when, at the age of five, my parents astutely perceived
    that God had made a mistake, and brought me to the eminent Dr. Charhack
    of Children's Hospital in Los Angeles. Dr. Charhack wrote the famous med-
    ical treatise "God broke it; I fixed it." After only three short years of surgery,
    my "hernia" was cured. I pray that I will have the means to repay, in some
    measure, the American Urological Association, for all they have done for by
    benefit. I am having some trouble, though, in connecting the timing mech-
    anism to the fuse. (Thomas in *Hermaphrodites With Attitude*, 1995: 16)

74. The ISNA has expressed concern at the uncritical acceptance of Victorian
    classifications by Anne Fausto-Sterling. "We call those ones pseudo-intersex-
    uals, ... because that is how we fool ourselves that the world is not full of
    intersexuals" (Chase in Burke, 1996: 225).
75. The "body part" criterion of sex is not limited to people with intersex condi-
    tions. Polycystic Ovarian Syndrome is characterized in women by infertility or
    higher risk miscarriage, hirsutism ("excess" hair growth), amenorrhea or
    oligomenorrhea (irregular or no menstrual cycle), menorrhagia ("excessive"
    menstrual bleeding), anovulation, androgenic alopecia (male pattern hair loss),
    and excess androgen production (Willmott, 2000). Women born with Mayer-
    Rokitansky-Kuster-Hauser Syndrome have a chromosome karyotype of 46XX,
    "normal" looking external genitalia, and complete absence of vagina, fallopian
    tubes, cervix, and uterus (Morris, 2000). Women with Mayer-Rokitansky-Kuster-
    Hauser Syndrome, some women with intersex conditions, and women with
    acute cases of vaginal atresia (absence or closure of a "normal" body orifice) also
    have reconstructive vaginal surgery (Whittle, 2000). Rather than attempting to
    create definitive boundaries of "sexual difference", the more pertinent task is to
    explore why certain body parts, such as vaginas, define "sex".
76. Many people with intersex conditions do not become aware of their condi-
    tion until adolescence. Still other individuals may never become aware that
    they have an intersex condition.

Chapter 8: How to Have Sex without Women or Men

77. Taking into account new-materialism does not in any way obviate arguments
    about the social construction of scientific knowledge, outlined in Chapter 1.
    For instance, Schiebinger outlines that the history of the study of bacteria
    was infused with *a priori* notions of "sex" and gender from the outset (1999).
    Until the 1940s, bacteria were assumed to be asexual. After that time, the
    "sex life" of bacteria were described in heterosexual terms. Specifically,
    bacteria were defined as "female" or "male" based on the absence or presence
    (respectively) of a "fertility" or F-factor (females are designated F−); (males
    are designated F+):

    To transfer genetic material, the "donor" or "male" extends its *sex pili* to the
    "recipient" or "female." Unlike the case in higher organisms, the chromosomal

transfer is unidirectional from male to female and the *male*, not the female, produces offspring. Futher when the F+ cell transfers a copy of its F− factor to an F− partner, the recipient becomes male or F+. Because the donor cell replicates its F+ factor during conjugation, it too remains F+. Thus all cells in mixed cultures rapidly become male (F+) donor cells: the females change into males, the males remain males, and everyone is happy. A recombinant F− (female) cell results only from a "disrupted" or failed transfer of DNA ... (1999: 149–150)

The importation of heteronomative ideology onto analyses of bacteria persisted until the 1990s, decelerating the recognition of alternative accounts of bacterial "sexual relations" such as the more obvious transsexual and transspecies interpretations.

Bibliography

Abbott, D.H. (1987) "Behaviorally Mediated Suppression of Reproduction in Female Primates," *Journal of Zoology*, 213: 455–70.

Abramson, P. and Pinkerton, S. (eds) (1995) *Sexual Nature, Sexual Culture*. Chicago, IL: University of Chicago Press.

Abramson, P. and Pinkerton, S. (1995) "Introduction: Nature, Nurture, and In-Between," in P. Abramson and S. Pinkerton (eds) *Sexual Nature, Sexual Culture*. Chicago, IL: University of Chicago Press, pp. 1–16.

Adler, T. (1997) " 'Animals' Fancies: Why members of Some Species Prefer their own Sex," *Science News Online*, April 1. http://www.sciencenews.org/sn_arc97/1_4_97/bob1.htm pp. 1–5. Accessed on May 7, 2003.

Ainley, M. (1990) *Despite the Odds*. Montreal: Véhicle Press.

Allen, C. (1997) "Inextricably Entwined: Politics, Biology, and Gender-Dimorphic Behavior," in P. Gowaty (ed.) *Feminism and Evolutionary Biology. Boundaries, Intersections, and Frontiers*. New York: Chapman and Hall, pp. 515–21.

Altmann, J. (1980) *Baboon Mothers and Infants*. Cambridge, MA.: Harvard University Press.

Bagemihl, B. (1999) *Biological Exuberance. Animal Homosexuality and Natural Diversity*. New York: St. Martin's Press, pp. 269–663.

Balsamo, A. (1996) *Technologies of the Gendered Body: Reading Cybrog Women*. Durham, NC and London: Duke University Press.

Barad, K. (1998) "Getting Real: Technoscientific Practices and the Materialization of Reality," *Differences: A Journal of Feminist Cultural Studies*, 10(2): 87–128.

Barad, K. (2001) "Scientific Literacy -> Agential Literacy = (Learning + Doing) Scientific Responsibility," in M. Mayberry, B. Subramaniam and L. Weasel (eds) *Feminist Science Studies*. New York: Routledge, pp. 226–46.

Barchilon, J. (1965) "Introduction," In M. Foucault (ed.) *Madness and Civilisation: A History of Insanity in the Age of Reason*. London: Tavistock, pp. v–viii.

Battersby, C., Constable, C., Jones, R., and Purdom, J. (eds) (2000) "Going Australian: Reconfiguring Feminism and Philosophy," *Hypatia*, 15(2): 1–216.

Benirschke, K. (1981) "Hermaphrodites, Freemartins, Mosaics, and Chimaeras in Animals," in C.R. Austin and R.G. Edwards (eds) *Mechanisms of Sex Differentiation in Animals and Man*, London: Academic Press, pp. 421–63.

Bernstein, I. (1997) "Females and Feminists, Science and Politics, Evolution and Change: An Essay," in P. Gowaty (ed.) *Feminism and Evolutionary Biology. Boundaries, Intersections, and Frontiers*. New York: Chapman and Hall, pp. 575–82.

Bird, G. *et al.* (1982) "Another Example of Haemopoietic (Twin) Chimaerism in a Subject Unaware of Being a Twin," *Journal of Immunogenetics*, 9: 317–22.

Birke, L. (1999) *Feminism and the Biological Body*. Edinburgh: Edinburgh University Press.

Birkhead, T.R. and Møller, A.P. (1993a) "Sexual Selection and the Temporal Separation of Reproductive Events: Sperm Storage Data from Reptiles, Birds, and Mammals," *Biological Journal of the Linnaean Society*, 50: 295–311.

Birkhead, T.R. and Møller, A.P. (1993b) "Female Control of Paternity," *Trends in Ecology and Evolution*, 8: 100–4.

Birkhead, T.R., Møller, A.P. and Sutherland, W.J. (1993) "Why do Females Make it so Difficult for Males to Fertilize Their Eggs," *Journal of Theoretical Biology*, 161: 51–60.

Bleier, R. (1984) *Science and Gender*. New York: Pergamon Press.

Bodeker, H. 1997. "Portrait of the Artist as a Young Herm": http://www.qis.net/~triea/hieke.html.

Bordo, S. (1993) *Unbearable Weight: Feminism, Western Culture, and the Body*. Berkeley and Los Angeles, CA: University of California Press.

Boswell, J. (1980) *Christianity, Social Tolerance and Homosexuality: Gay People in Western Europe from the Beginning of the Christian Era to the Fourteenth Century*. Chicago, IL: University of Chicago Press.

Braidotti, R. (1994) *Nomadic Subjects: Embodiment and Sexual Difference in Contemporary Feminist Theory*. New York: Columbia University Press.

Braidotti, R. (2000) "Teratologies," in I. Buchanan and C. Colebrook (eds) *Deleuze and Feminist Theory*, Edinburgh: Edinburgh University Press, pp. 156–72.

Braidotti, R. and Lykke, N. (1996) *Between Monsters, Goddesses and Cyborgs*. New York: Zed Books.

Bray, A. and Colebrook, C. (1998) "The Haunted Flesh: Corporeal Feminism and the Politics of (Dis)Embodiment," *Signs: Journal of Women in Culture and Society*, 24(1): 35–67.

Broadhurst Dixon, J. and Cassidy, E. (1998) *Virtual Futures*. London: Routledge.

Brockman, J. and Matson, K. (1996) *How Things Are. A Science Tool-kit for the Mind*. London: Phoenix.

Brown, N. (1999a) "Debates in Xenotransplantation: On the Consequences of Contradiction," *New Genetics and Society*, 18(2–3): 181–96.

Brown, N. (1999b) "Xenotransplantation: Normalizing Disgust," *Science as Culture*, 8(3): 327–55.

Brown, N. and Michael, M. (2001) "Transgenics, Uncertainty and Public Credibility," *Transgenic Research*, 10: 279–83.

Brown, T.M. (1985) "Descartes, Dualism, and Psychosomatic Medicine," in W.F. Bynum, R. Porter and M. Shepherd (eds) *The Anatomy of Madness, Essays in the History of Psychiatry*. London and New York: Tavistock, pp. 40–62.

Brown, W. (1996) "The Impossibility of Women's Studies," *Differences: A Journal of Feminist Cultural Studies*, 9(3): 79–101.

Buchanan, I. and Colebrook, C. (eds) (2000) *Deleuze and Feminist Theory*. Edinburgh: Edinburgh University Press.

Bullough, V. (1974) "Transvestites in the Middle Ages," *American Journal of Sociology*, 79: 1381–9.

Burke, P. (1996) *Gender Shock: Exploding the Myths of Male and Female*. New York: Anchor Books.

Butler, J. (1990) *Gender Trouble: Feminism and the Subversion of Identity*. New York: Routledge.

Butler, J. (1993) *Bodies That Matter: On the Discursive Limits of 'Sex'*. New York: Routledge.

Butler, J. (1997) *The Psychic Life of Power*. Stanford: Stanford University Press.

Cann, R., Stoneking, M., and Wilson, A. (1987) "Mitochondrial DNA and Human Evolution," *Nature*, 325: 31–6.

Capra, F. (1975) *The Tao of Physics*. Berkeley, CA: Shambhala.

Charnov, E. (1982) *Sex Allocation*. Princeton, NJ: Princeton University Press.

Chase, C. (1998) "Hermaphrodites with Attitude", *GLQ: A Journal of Lesbian and Gay Studies*, 4(2): 189–212.

Cheah, P. (1996) "Mattering," *Diacritics*, 26(1): 108–39.

Cheek, A.O. and McLachlan, J. (1998) "Environmental Hormones and the Male Reproductive System," *Journal of Andrology*, 19(1): 5–10.

Choudhury, S. (1995) "Divorce in Birds: A Review of the Hypotheses," *Animal Behavior*, 50: 413–29.

Clark, E., Norris, D., and Jones, R. (1998) "Interactions of Gonadal Steroids and Pesticides (DDT, DDE) on Gonaduct Growth in Larval Tiger Salamanders," *General and Comparative Endocrinology*, 109: 94–105.

Clark, N. (2000)" "Botonizing the Ashphalt?" The Complex Life of Cosmopolitan Bodies," *Body and Society*, 6(3–4): 23–30.

Clark, N. (2001) "De/feral: Introduced Species and the Metaphysics of Conservation," in L. Simmons and H. Worth (eds) *Derrida Downunder*. Palmerston North: Dunmore Press, pp. 86–106.

Clarke, A. (2000) *Greetings, Carbon-based Bipeds!* London: HarperCollins.

Cohen, P. (1995) "The Boy Whose Blood Has No Father," *New Scientist*, 7 October, p. 16.

Colborn, T., Dumanoski, D., and Myers, J.P. (1996) *Our Stolen Future*. London: Little Brown and Company.

Colebrook, C. (2000a) "Is Sexual Difference a Problem?," in I. Buchanan and C. Colebrook (eds) *Deleuze and Feminist Theory*. Edinburgh: Edinburgh University Press, pp. 110–27.

Colebrook, C. (2000b) "From Radical Representations to Corporeal Becomings: The Feminist Philosophy of Lloyd, Grosz and Gatens," *Hypatia*, 15(2): 76–93.

Collins, R.J. (1994) "Artificial Evolution and the Paradox of Sex," in R. Parton (ed.) *Computing with Biological Metaphors*. London: Chapman and Hall, pp. 244–63.

Connell, R.W. (1996) "New Directions in Gender Theory, Masculinity Research and Gender Politics," *Ethnos*, 19: 157–76.

Coventry, M. (1999) "Finding the Words" in Domurant Dreger, A. (ed.) *Intersex in the Age of Ethics*. Hagerstown, Maryland: University Publishing Group, pp. 71–8.

Dagg, A.I. (1984) "Homosexual Behavior and Female-Male Mounting in Mammals," *Mammal Review*, 14: 155–85.

Daly, M. (1978) "The Cost of Mating," *American Naturalist*, 112: 771–4.

Darby, T. and Emberley, P. (1996) ' "Political Correctness' and the Constitution," in A. Peacock (ed.) *Rethinking the Constitution: Perspectives on Canadian Constitutional Reform, Interpretation and Theory*. Oxford: Oxford University Press, pp. 233–48.

Darwin, C. (1859/1998) *The Origin of Species*. Ware, Hertfordshire: Wordsworth Editions Limited.

Daston, L. and Park, K. (1998) *Wonders and the Order of Nature*. New York: Zone Books.

Davis, M. (1998) *Ecology of Fear: Los Angeles and the Imagination of Disaster*. New York: Metropolitan Books.

Dawkins, R. (1989) *The Selfish Gene*. New Edition. Oxford and New York: Oxford University Press.

DeBeauvoir, S. (1949/76) *The Second Sex*. Harmondsworth: Penguin.

Deichmann, U. (1996) "Konrad Lorenz, Ethology, and Natural Socialist Racial Doctrine," *Biologists Under Hitler*. Cambridge, Mass.: Harvard University Press, pp. 179–205.

De Landa, M. (1991) *War in the Age of Intelligent Machines*. New York: Swerve Editions.

De Landa, M. (1995) "Uniformity and Variability: An Essay in the Philosophy of Matter," *Doors of Perception 3 Conference*, pp. 1–8.

De Landa, M. (1997a) "Immanence and Transcendence in the Genesis of Form," *The South Atlantic Quarterly*, 96(3): 499–514.

De Landa, M. (1997b) *A Thousand Years of Nonlinear History*. New York: Swerve Editions.

De Landa, M. (2000) "Deleuze and the Open-ended Becoming of the World" www.brown.edu/ Departments/Watson _Institute/programs/gs/VirtualY2K/De Landa.html, pp. 1–10.

Deleuze, G. (1994) *Difference and Repetition*. New York: Columbia University Press.

Deleuze, G. and Guattari, F. (1983) *Anti-Oedipus. Capitalism and Schizophrenia*. Minneapolis, MN: University of Minnesota Press.

Deleuze, G. and Guattari, F. (1987) *A Thousand Plateaus*. London: Athlone Press.

Delphy, C. (1984) *Close to Home*. London: Hutchinson.

Delphy, C. (1994) "Changing Women in a Changing Europe," *Women's Studies International Forum*, 17: 187–201.

Denniston, R.H. (1980) "Ambisexuality in Animals," in J. Marmor (ed.) *Homosexual Behavior: A Modern Reappraisal*. New York: Basic Books, pp. 35–40.

Derrida, J. (1978) *Writing and Difference*. Trans. A. Bass. London and Henley: University of Chicago Press.

deWaal, F. (1995) "Sex as an Alternative to Aggression in the Bonobo," in P. Abramson and S. Pinkerton (eds) *Sexual Nature, Sexual Culture*. Chicago, IL: University of Chicago Press, pp. 37–56.

Dewhurst, C. and Gordon, R. (1969) *The Intersexual Disorders*. London: Balliere, Tindall and Cassell.

Dickemann, M. (1979) "Female infanticide and reproductive strategies of stratified human societies," in N. Chagnon and W. Irons (eds) *Evolutionary Biology and Human Social Behavior*. North Scituate, MA: Duxbury.

Diprose, R. (1991) "A 'Genetics' That Makes Sense," in R. Diprose and R. Ferrell (eds) *Cartographies: Poststructuralism and the Mapping of Bodies and Spaces*. Sydney: Allen and Unwin, pp. 65–76.

Doane, M.A. (1997) "Technology and Sexual Difference: Apocalyptic Scenarios at Two 'Fins de Siècle,' *Differences: A Journal of Feminist Cultural Studies*, 9(2): 1–24.

Dobzhansky, T. (1956) "What is an Adaptive Trait?" *The American Naturalist*, 190: 337–47.

Dolk *et al*. (1998) "Risk of Congenital Anomalies Near Hazardous Waste Landfill Sites in Europe: The EUROHAZCON Study," *The Lancet*, 352, August 8, pp. 423–27.

Donchin, A. (1989) "The Growing Feminist Debate over the New Reproductive Tehcnologies," *Hypatia*, 4(3): 136–49.

Dreger, A. (1998) *Hermaphrodites and the Medical Invention of Sex*. Cambridge, MA: Harvard University Press.

Dunbar, R.I.M. (1980) "Determinants and Evolutionary Consequences of Dominance among Female Gelada Baboons," *Behavior, Ecology and Sociobiology*, 7: 253–65.

Durkheim, E. (1970) *Suicide*. London: Routledge and Kegan Paul.

Dynes, W.R. (1987) "Animal Homosexuality," in *Homosexuality: A Research Guide*. New York and London: Garland Publishing, pp. 743–9.

Ebensperger, L.A. and Tamarin, R.H. (1997) "Use of Fluorescent Powder to Infer Mating Activity of Male Rodents," *Journal of Mammalogy*, 78: 888–93.

Ehrhardt, A.A. 1985. "Sexual Orientation After Prenatal Exposure To Exogenous Estrogen," *Archives of Sexual Behaviour*, 14: 57–77.

Elekonich, M. (2001) "Contesting Territories: Female–Female Aggression and Song Sparrows," in M. Mayberry, B. Subramaniam, and L. Weasel (eds) *Feminist Science Studies*. New York: Routledge, pp. 97–105.

Epstein, J. (1990) "Either/Or–Neither/Both: Sexual Ambiguity and the Ideology of Gender," *Genders* 7: 19–25.

Estes, R.D. (1991) "The Significance of Horns and Other Male Secondary Sexual Characteristics in Female Bovids," *Applied Animal Behavior Science*, 29: 403–51.

Fausto-Sterling, A. (1992) *Myths of Gender*. New York: HarperCollins.

Fausto-Sterling, A. (1993) "The Five Sexes: Why Male and Female Are Not Enough," *The Sciences*, March–April: 20–25.

Fausto-Sterling, A. (1997) "Feminism and Behavioral Evolution: A Taxonomy," in P. Gowaty (ed.) *Feminism and Evolutionary Biolology. Boundaries, Intersections, and Frontiers*. New York: Chapman and Hall, pp. 42–60.

Fausto-Sterling, A. (2000) *Sexing the Body*. New York: Basic Books.

Featherstone, M. and Burrows, R. (1995) *Cyberspace, Cyberbodies, Cyberpunk*. London: Sage.

Fedigan, L. (1984) "Sex Ratios and Sex Differences in Primatology," *American Journal of Primatology*, 7: 305–8.

Fedigan, L. (1991) *Primate Paradigms: Sex Roles and Social Bonds*. Chicago, IL: University of Chicago Press.

Feinberg, L. (1996) *Transgender Warriors*. Boston, MA: Beacon Press.

Ferguson, H. (1997) "Me and My Shadows: On the Accumulation of Body-Images in Western Society Part Two — The Corporeal Forms of Modernity," *Body and Society*, 3(3): 1–31.

Ferrière, R. and Fox, G.A. (1995) "Chaos and Evolution," *Trends in Ecology and Evolution*, 10: 480–5.

Flax, J. (1990) "The End of Innocence," in L. Nicholson (ed.) *Feminism/ Postmodernism*. London: Routledge, pp. 39–62.

Flieger, J. (2000) "Become-Woman: Deleuze, Schreber and Molecular Identification," in I. Buchanan and C. Colebrook (eds) *Deleuze and Feminist Theory*. Edinburgh: Edinburgh University Press, pp. 38–63.

Foucault, M. (1965) *Madness and Civilisation: A History of Insanity in the Age of Reason*. London: Tavistock.

Foucault, M. (1979) *The History of Sexuality: An Introduction*. Vol. 1. New York: Vintage.

Foucault, M. (1980) *Herculine Barbin: Being the Recently Discovered Memoris of a 19th century French Intersexual*. New York: Pantheon.

Foucault, M. (1994a) *The Birth of the Clinic. An Archaeology of Medical Perception*. London: Vintage Books.

Foucault, M. (1994b) *The Order of Things. An Archaeology of the Human Sciences*. New York: Vintage Books.

Foucault, M. (1997) "The abnormals," in P. Rabinow (ed.) *Michel Foucault: Ethics*. New York: The New Press, pp. 51–8.

Franklin, S. (1995) "Romancing the Helix: Nature and Scientific Discovery," in L. Pearce and J. Stacey (eds) *Romance Revisited*. London: Lawrence and Wishart.

Franklin, S. (1997) *Embodied Progress. A Cultural Account of Assisted Conception*. London and New York: Routledge.

Franklin, S. (2000) "Life Itself," in S. Franklin, C. Lury, and J. Stacey (eds) *Global Nature, Global Culture*. London: Sage, pp. 188–227.

Franklin, S. (2001) "Biologization Revisited: Kinship Theory in the Context of the New Biologies," in Franklin, S. and S. McKinnon (eds) *Relative Values. Reconfiguring Kinship Studies*. Durham, NC: Duke University Press, pp. 302–25.

Franklin, S. and McKinnon, S. (eds) (2001) *Relative Values. Reconfiguring Kinship Studies*. Durham, NC: Duke University Press.

Franklin, S., Lury, C., and Stacey, J. (2000) *Global Nature, Global Culture*. London: Sage.

Fraser, M. (1999/2000) "Creative Affects," *New Formations*, 39: 55–69.

Fuss, D. (1989) *Essentially Speaking*. New York and London: Routledge.

Futuyama, D.J. (1979) *Evolutionary Biology*. Sunderland, MA.: Sinauer.

Futuyama, D.J. and Risch, S.J. (1984) "Sexual Orientation, Sociobiology, and Evolution," *Journal of Homosexuality*, 9: 157–68.

Gadpaille, W.J. (1980) "Cross-species and cross-cultural Contributions to Understanding Homosexual Activity," *Archives of General Psychiatry*, 37: 349–56.

Gallup, G. and Suarez, S. (1983) "Homosexuality as By-product of Selection for Optimal Heterosexual Strategies," *Perspectives in Biological Medicine*, 26: 315–21.

Garfinkel, H. and Stoller, R. (1967). "Passing and the Managed Achievement of Sex Status in an 'Intersexed' Person. Part I" in *Studies in Ethnomethodology*. Cambridge: Polity Press, pp. 116–185.

Gartrell, N. (1982) "Hormones and Homosexuality," in W. Paul, J.D. Weinrich, J.C. Gonsiorek, and M.E. Hotvedt (eds) *Homosexuality. Social, Psychological and Biological Issues*. London: Sage Publications.

Gatens, M. (2000) "Feminism as 'Password': Re-thinking the 'Possible' with Spinoza and Deleuze," *Hypatia*, 15(2): 59–75.

Gelbart, W. (1998) "Data Bases in Genomic Research," *Science*, 282: 660.

Golden, C. (2000) "Still Seeing Differently, After All These Years," *Feminism and Psychology*, 10: 30–5.

Golden, R. *et al.* (1998) "Environmental Endocrine Modulators and Human Health: An Assessment of the Biological Evidence," *Critical Review of Toxicology*, 28(2): 109–227.

Gould, S.J. (1981) *The Mismeasure of Man*. New York: Morton Press.

Gould, S.J. (1987) "Freudian Slip," *Natural History*, 96: 14–21.

Gould, S.J. (1991) "Exaptation: A Crucial Tool for an Evolutionary Psychology," *Journal of Social Issues*, 47: 43–65.

Gould, S.J. (1994) "Curveball. Book Review of The Bell Curve: Intelligence and Class Structure in American Life," *New Yorker*, 70: 139–49.

Gould, S.J. (2000) *Wonderful Life. The Burgess Shale and the Nature of History*. London: Vintage.

Gould, S.J. (2002) *The Structure of Evolutionary Theory*. Cambridge, MA. and London: The Belknap Press of Harvard University Press.

Gould, S.J. and Lewontin, R.C. (1979) "The Spandrels of San Marcos," *Proceedings of the Royal Society of London*, 205: 581–98.

Gowaty, P.A. (1982) "Sexual Terms in Sociobiology: Emotionally Evocative and Paradoxically, Jargon," *Animal Behavior*, 30: 630–1.

Gowaty, P.A. (1997a) *Feminism and Evolutionary Biolology. Boundaries, Intersections, and Frontiers*. New York: Chapman and Hall.

Gowaty, P.A. (1997b) "Introduction: Darwinian Feminists and Feminist Evolutionists," in P. Gowaty (ed.) *Feminism and Evolutionary Biolology. Boundaries, Intersections, and Frontiers*. New York: Chapman and Hall, pp. 1–17.

Gowaty, P.A. (1997c) " 'Principles of Females' Perspectives in Avian Behavioral Ecology," *Journal of Avian Biology*, 28: 95–102.

Gray, C. (1995) *The Cyborg Handbook*. New York: Routledge.

Gray, R. (1997) " 'In the Belly of the Monster': Feminism, Developmental Systems, and Evolutionary Explanations," in P. Gowaty (ed.) *Feminism and Evolutionary Biology. Boundaries, Intersections, and Frontiers*. New York: Chapman and Hall, pp. 385–414.

Griffin, J.E. and Wilson, J.D. (1992) "Disorders of Sexual Differentiation," in P.C. Walsh, A.B. Retik, T.A. Stamey, and E.D. Vaughan (eds) *Campbells Urology*. Philadelphia, PA: Saunders, pp. 1509–37.

Grosz, E. (1986) "Derrida and the Limits of Philosophy," *Thesis Eleven*, 14: 26–43.

Grosz, E. (1995) "Animal Sex. Libido as Desire and Death," in E. Grosz and E. Probyn (eds) *Sexy Bodies*. London: Routledge, pp. 278–99.

Grosz, E. (1999a) *Becomings. Explorations in Time, Memory, and Future*. Ithaca, NY and New York: Cornell University Press.

Grosz, E. (1999b) "Darwin and Feminism: Preliminary Investigations for a Possible Alliance," *Australian Feminist Studies*, 14(29): 31–45.

Grosz, E. (1999c) "Thinking the New: Of Futures Yet Unthought," in E. Grosz (ed.) *Becomings. Explorations in Time, Memory, and Futures*. Ithaca, NY and London: Cornell University Press.

Haldane, J.B.S. (1928) *Possible Worlds and Other Papers*. New York: Harper and Brothers.

Hale, C.J. (1997) "Leatherdyke boys and Their Daddies: How to Have Sex Without Women or Men," *Social Text 52/53*, 15(3–4): 225–38.

Halperin, D. (1990) *One Hundred Years of Homosexuality and Other Essays on Greek Love*. New York: Routledge.

Hankinson-Nelson, L. and Nelson, J. (1996) *Feminism, Science and the Philosophy of Science*. Dordrecht: Kluwer.

Haraway, D. (1989) *Primate Visions. Gender, Race and Nature in the World of Modern Science*. New York: Routledge.

Haraway, D. (1991) "A Cyborg Manifesto: Science, Technology, and Socialist-Feminism in the Late Twentieth Century," in *Simians, Cyborgs and Women*. New York: Routledge, pp. 149–82.

Haraway, D. (1992) "When Man™ is on the Menu," in J. Crary and S. Kwinter (eds) *Incorporations*. New York: Urzone Books, pp. 38–44.

Haraway, D. (1997) *Modest_Witness@Second_Millennium. FemaleMan©_Meets_OncoMouse™*. New York and London: Routledge.

Haraway, D. (2001) "More than Metaphor," in M. Mayberry, B. Subramaniam, and L. Weasel (eds) *Feminist Science Studies*. New York: Routledge, pp. 81–6.

Harding, J. (1996) "Sex and Control: The Hormonal Body," *Body and Society*, 2(1): 99–111.

Harding, S. (1986) *The Science Question in Feminism*. Ithaca, NY Cornell University Press.

Harding, S. (1991) *Whose Science? Whose Knowledge? Thinking from Women's Lives*. Ithaca, NY: Cornell University Press.

Hausman, B. (1995) *Changing Sex: Transsexualism,Technology and the Idea of Gender*. Durham: Duke University Press.

Hayles, N.K. (1990) *Chaos Unbound: Orderly Disorder in Contemporary Literature and Science*. Ithaca, NY: Cornell University Press.

Henrion, C. (1997) *Women in Mathematics: The Addition of Difference*. Bloomington, IN: Indiana University Press.

Hermaphrodites with Attitude (1995) Quaterly Publication of the Intersex Society of North America. http://www.isna.org.

Hird, M.J. (2000) "Gender's Nature: Intersexuals, Transsexualism and the 'Sex'/'Gender' Binary," *Feminist Theory*, 1(3): 347–64.

Hird, M.J. (2002) "Re(pro)ducing Sexual Difference," *Parallax*, 8(4): 94–107.

Hird, M.J. (2003a) "From the Culture of Matter to the Matter of Culture: Feminist Explorations of Nature and Science," *Sociological Research Online*, 8(1). http://www.socresonline.org.uk/8/1/hird.html

Hird, M.J. (2003b) "Considerations for a Psycho-analytic Theory of Gender Identity and Sexual Desire: The Case of Intersex," *Signs: Journal of Women in Culture and Society*, 28(4): 1067–92.

Hird, M.J. (2003c) "New Feminist Sociological Directions," *Canadian Journal of Sociology*, 28(4): 447–62.

Hird, M.J. (2004a) "Naturally Queer," *Feminist Theory*, 5(1): 85–9.

Hird, M.J. (2004b) "Chimerism, Mosaicism and the Cultural Construction of Kinship," *Sexualities*, 7(2): 225–40.

Hird, M.J. and Germon, J. (2001) "The Intersexual Body and the Medical Regulation of Gender," in K. Backett-Milburn and L. McKie (eds) *Constructing Gendered Bodies*. London: Palgrave, pp. 162–78.

Ho, M.-W., Saunders, P., and Fox, S. (1986) "A New Paradigm for Evolution," *New Scientist*, 109(1497): 41–3.

Holmes, M. (1995) "Queer Cut Bodies: Intersexuality and Homophoba in Medical Practice." http://cwis.use.edu/Library/QF/queer/papers/holmes.long.html.

Hood-Williams, J. (1996) "Goodbye to Sex and Gender," *The Sociological Review*, 44(1): 1–16.

Hrdy, S.B. (1974) "Male-male Competition and Infanticide among the Langurs (*Presbytis entellus*) of Abus, Rajasthan," *Folia Primatologica*, 22: 19–58.

Hrdy, S.B. (1981) *The Woman that Never Evolved*. Cambridge, MA.: Harvard University Press.

Hrdy, S.B. (1986) "Empathy, Polyandry, and the Myth of the Coy Female," in R. Bleier (ed.) *Feminist Approaches to Science*. New York: Pergamon Press, pp. 119–46.

Hrdy, S.B. (1997) "Rasing Darwin's consciousness: Female Sexuality and the Prehominid Origins of Patriarchy," *Human Nature*, 8: 1–49.

Hubbard, R. (1979) "Have Only Men Evolved?" in R. Hubbard, M. Henifin, and B. Fried (eds) *Women Look at Biology Looking at Women. A Collection of Feminist Critiques*. Cambridge, MA: Schenkman Publishing, pp. 7–35.

Hubbard, R. (1989) "Science, Facts, and Feminism," in N. Tuana (ed.) *Feminism and Science*. Bloomington, IN: Indiana University Press, pp. 119–31.

Hubbard, R. and Wald, E. (1993) *Exploding the Gene Myth*. Boston, MA: Beacon Press.

Hubbard, R. Henifin, M., and Fried, B. (eds) (1979) *Women Look at Biology Looking at Women. A Collection of Feminist Critiques*. Cambridge, MA: Schenkman Publishing.

Hutchinson, G. (1959) "A Speculative Consideration of Certain Possible Forms of Sexual Selection in Man," *American Naturalist*, 93(869): 81–91.

Irigaray, L. (1985) *This Sex Which Is Not One*. Ithaca, NY: Cornell University Press.

Jackson, J. (2001) "Unequal Partners: Rethinking Gender Roles in Animal Behavior," in M. Mayberry, B. Subramaniam, and L. Weasel (eds) *Feminist Science Studies*. New York: Routledge, pp. 115–19.

Jonson, A. (1999) "Still Platonic After All These Years: Artificial Life and Form/Matter Dualism," *Australian Feminist Studies*, 14(29): 47–61.

Kauffman, S. (1993) *The Origins of Order*. New York: Oxford University Press.

Keane, H. (1999) "Adventures of the Addicted Brain," *Australian Feminist Studies*, 14(29): 63–76.

Keller, E.F. (1977) "The Anomaly of a Woman in Physics," in S. Ruddick and P. Daniels (eds) *Working It Out*. New York: Pantheon Books.

Keller, E.F. (1983) *A Feeling for the Organism*. New York: W.H. Freeman and Company.

Keller, E.F. (1989) "The Gender/Science System: Or, is Sex to Gender as Nature is to Science?," in N. Tuana (ed.) *Feminism and Science*. Bloomington, IN: Indiana University Press, pp. 33–44.

Keller, E.F. (2000) *The Century of the Gene*. Cambridge, MA: Harvard University Press.

Kelly, K. (1994) *Out of Control. The New Biology of Machines*. London: Fourth Estate.

Kerin, J. (1999) "The Matter at Hand: Butler, Ontology and the Natural Sciences," *Australian Feminist Studies*, 14(29): 91–104.

Kessler, S. (1990) "The Medical Construction of Gender: Case Management of Intersexed Infants," *Signs: Journal of Women in Culture and Society*, 16: 3–26.

Kessler, S. (1998) *Lessons From the Intersexed*. New Brunswick, NJ: Rutgers University Press.

Kinsman, S. (2001) "Life, Sex and Cells," in M. Mayberry, B. Subramaniam, and L. Weasel (eds) *Feminist Science Studies*. New York: Routledge, pp. 193–203.

Kirby, V. (1997) *Telling Flesh. The Substance of the Corporeal*. New York: Routledge.

Kirby, V. (1999) "Human Nature," *Australian Feminist Studies*, 14(29): 19–29.

Kirby, (2001) "Quantum Anthropologies," in L. Simmons and H. Worth (eds) *Derrida Downunder*. Palmerston North: Dunmore Press, pp. 53–68.

Kirsch, J. and Rodman, J. (1982) "Selection and Sexuality. The Darwinian View of Homosexuality," in W. Paul, J.D. Weinrich, J.C. Gonsiorek, and M.E. Hotvedt (eds) *Homosexuality. Social, Psychological and Biological Issues*. London: Sage Publications.

Kitcher, P. (1985) *Vaulting Ambition: Sociobiology and the Quest for Human Nature*. Cambridge, MA: MIT Press.

Kleinman, D.G. (1977) "Monogamy in Mammals," *Quarterly Review of Biology*, 52: 39–69.

Kohlstedt, S. and Longino, H. (1997) *Women, Gender and Science: New Directions*. Chicago, IL: University of Chicago Press.

Komers, P.E. (1997) "Behavioral Plasticity in Variable Environments," *Canadian Journal of Zoology*, 75: 161–9.

Krizek, G.O. (1992) "Unusual Interaction Between a Butterfly and a Beetle: 'Sexual Paraphilia' in Insects?," *Tropical Lepidoptera*, 3(2): 118.

Kuroda, S. (1980) "Social Behavior of the Pygmy Chimpanzee," *Primates*, 21: 181–97.

Labov, J.B. (1981) "Pregnancy Blocking in Rodents: Adaptive Advantages for Females," *American Naturalist*, 118: 361–71.

Laidman, J. (2000) "Reproduction a Touch-and-go Thing for Fungus," *Nature*, July 24: 1–3.

Lamarck, J.B. (2004) *Lamarck's Open Mind: The Lectures*. High Sierra Books.

Lancaster, J. (1989) "Women in Biosocial Perspective," in S. Morgan (ed.) *Gender and Anthropology*. Washington, DC: American Anthropological Association, pp. 95–115.

Lancaster, J. (1991) "A Feminist and Evolutionary Biologist Looks at Women," *Yearbook of Physical Anthropology*, 34: 1–11.

Landrigan, P. *et al*. (1998) "Children's Health and the Environment: A New Agenda for Prevention Research," *Environmental Health Perspectives*, 106(3): 787–94.

Laqueur, T. (1990) *Making Sex*. Cambridge, MA: Harvard University Press.

Latour, B. (2000a) *We Have Never Been Modern*. Hemel Hempstead: Harvester Wheatsheaf.

Latour, B. (2000b) "When Things Strike Back: A Possible Contribution of 'Science Studies' to the Social Sciences," *British Journal of Sociology*, 51(1): 107–23.

Lawton, M., Garstka, W., and Hanks, J.C. (1997) "The Mask of Theory and the Face of Nature," in P. Gowaty (ed.) *Feminism and Evolutionary Biolology. Boundaries, Intersections, and Frontiers*. New York: Chapman and Hall, pp. 63–85.

Lee, P., Mazur, T., Danish, R., Amrhein, J., Blizzard, R., Money, J. and Migeon, C. (1980) "Micropenis. I. Criteria, Etiologies and Classification," *The John Hopkins Medical Journal*, 146: 156–63.

LeVay, S. (1991) "A Difference in Hypothalamic Structure Between Heterosexual and Homosexual Men," *Science*, 253: 1034–1037.

LeVay, S. (1993) *The Sexual Brain*. Cambridge, MA: The MIT Press.

LeVay, S. (1996) *Queer Science. The Use and Abuse of Research into Homosexuality*. Cambridge, MA: The MIT Press.

Lingus, A. (1994) *Foreign Bodies*. New York: Routledge.

Lock, M. (1997) "Decentering the Natural Body: Making Difference Matter," *Configurations*, 5: 267–92.

Longino, H. (1990) *Science as Social Knowledge*. Princeton, NJ: Princeton University Press.

Lorber, J. (1994) *Paradoxes of Gender*. New Haven, CT: Yale University Press.

Lottring, S. (1989) *Foucault Live: Collected Interviews, 1961–1984*. New York: Semiotext(e).

Mackay, J. (2001) "Why Have Sex?" *British Medical Journal*, pp. 322–623.

Mackenzie, A. (1999) "Technical Materializations and the Politics of Radical Contingency," *Australian Feminist Studies*, 14(29): 105–18.

Malinowski, B. (1913) *The Family Among Australian Aborigines*. London: University of London Press.

Margulis, L. and Sagan, D. (1986) *Origins of Sex. Three Billion Years of Genetic Recombination*. New Haven, CT: Yale University Press.

Margulis, L. and Sagan, D. (1991) *Mystery Dance*. New York: Summit Books.

Margulis, L. and Sagan, D. (1995) *What is Life?* Berkeley, CA: University of California Press.

Margulis, L. and Sagan, D. (1997) *What is Sex?* New York: Simon and Schuster.

Marks, J. (2001) " 'We're Going to Tell These People Who They Really Are': Science and Relatedness," in S. Franklin and S. McKinnon (eds) *Relative Values. Reconfiguring Kinship Studies*. Durham, NC: Duke University Press, pp. 355–83.

Marshall, B. (2000) *Configuring Gender. Explorations in Theory and Politics*. Peterborough, Ontario: Broadview Press.

Marshall, B. (2001) " 'The Nature of the Body': Heterogendered Bodies and Sexual Medicine." *Paper presented at the British Sociological Association Conference*, University of Manchester, 9–12 April.

Martin, E. (1991) "The Egg and the Sperm: How Science Has Constructed a Romance Based on Stereotypical Male-Female Roles," *Signs: Journal of Women in Culture and Society*, 16(3): 485–501.

Martini, F. and Bartholomew, E. (2000) *Essentials of Anatomy and Physiology*. 2nd Edition. New Jersey: Prentice Hall.

Marx, K. (1887) *Capital: A Critique of Political Economy. Volume I: Capitalist Production*. London: Lawrence and Wishart.

May, R.M. (1988) 'How Many Species Are There?', *Science*, 241: 1441–9.

May, R. (1989) "The Chaotic Rhythms of Life," *New Scientist*, 124(1691): 37–41.

Mayberry, M., Subramaniam, B. and Weasel, L. (2001) *Feminist Science Studies*. New York: Routledge.

McKnight, J. (1997) *Straight Science? Homosexuality, Evolution and Adaptation*. London: Routledge.

McVean, G. and Hurst, L.D. (1996) "Genetic Conflicts and the Paradox of Sex Determination: Three Paths to the Evolution of Female Intersexuality in a Mammal," *Journal of Theoretical Biology*, 179: 199–211.

McWilliam and O'Donnell, S. (1998) "Probing Protocols", *Body and Society*, 4(3): 85–102.

Mead, G.H. (1934) *Mind, Self and Society*. Chicago, IL: The University of Chicago Press.

Meilwee, J. and Robinson, J. (1992) *Women in Engineering: Gender, Power, and Workplace Culture*. New York: SUNY Press.

Merrick, J. (1988) "Royal Bees: The Gender Politics of the Beehive in Early Modern Europe," *Studies in Eighteenth-Century Culture*, 18: 7–37.

Michael, M. and Carter, S. (2001) "The Facts About Fictions and Vice Versa: Public Understanding of Human Genetics," *Science as Culture*, 10(1): 5–32.

Milius, S. (1998) "When Birds Divorce: Who Splits, Who Benefits, and Who Gets the Next", *Science News*, 153: 153–5.

Miller, J. (1993) *The Passion of Michel Foucault*. New York: Simon and Schuster.

Moi, T. (1986) *The Kristeva Reader*. Oxford: Basil Blackwell.

Money, J. (1985) "The Conceptual Neutering of Gender and the Criminalisation of Sex," *Archives of Sexual Behaviour* 14: 279–1.

Money, J. (1994) Sex Errors of the Body. Baltimore: The John Hopkins University Press.

Money, J. and Ehrhardt, A.A. (1972) *Man and Woman Boy and Girl: The Differentiation and Dimorphism of Gender Identity from Conception to Maturity.* Baltimore, MD: Johns Hopkins University Press.

Money, J. and Tucker, P. (1976). Sexual Signatures on Being a Man or a Woman. London: Harrap.

Moore, K. and Persaud, T. (1998) *Before We Are Born. Essentials of Embryology and Birth Defects.* 5th Edition. Philadelphia, PA: W.B. Saunders Company.

Morris, E. (2000) *An Additional Monologue.* http://homestead.juno.com/mrkh1/files/AdditionalMonologue.htm.

Muma, K. and Weatherhead, P.J. (1989) "Male Traits Expressed in Females: Direct or Indirect Sexual Selection," *Behavioral Ecology and Sociobiology,* 25: 23–31.

Mundinger, P.C. (1980) "Animal Cultures and a General Theory of Cultural Evolution," *Ethology and Sociobiology,* 1: 83–223.

Murphy, J. (1989) "Is Pregnancy Necessary? Feminist Concerns about Ectogenesis", *Hypatia,* 4(3): 66–84.

Murphy, T. (1997) *Gay Science. The Ethics of Sexual Orientation Research.* New York: Columbia University Press.

Munster, A. (1999) "Is there Postlife after Postfeminism? Tropes of Technics and Life in Cyberfeminism," *Australian Feminist Studies,* 14(29): 119–30.

Nakashima, Y., Kuwamura, T., and Yogo, Y. (1995) "Why Be a Both-Ways Sex Changer," *Ethology,* 101: 301–7.

Nederman, D. and True, J. (1996) "The Third Sex: The Idea of the Hermaphrodite in Twelfth-century Europe," *Journal of the History of Sexuality,* 6(4): 497–517.

Nelson, J. (2002) "Microchimerism: Incidental Byproduct of Pregnancy or Active Participant in Human Health?" *Trends in Molecular Medicine,* 8(3): 109–13.

Neng, Y., Kruskall, M., Yunis, J., Knoll, J., Uhl, L., Alosco, S., Ohashi, M., Clavijo, O., Husain, Z., Yunis, E., Yunis, J., and Yunis, E. (2002) "Disputed Maternity Leading to Identification of Tetragametic Chimerism," *New England Journal of Medicine,* 346(20): 1545–52.

Nietzsche, F. (1974) *The Gay Science.* New York: Vintage Books.

Novas, C. and Rose, N. (2000) "Genetic Risk and the Birth of the Somatic Individual", *Economy and Society,* 29(4): 485–513.

Oberfield, S., Mondok, A., Shahrivar, F., Klein, J., and Levine, L. (1989) "Clitoral Size in Full-term Infants," *American Journal of Perinatology,* 6(4): 453–4.

Olkowski, D. (2000) "The End of Phenomenology: Bergon's Interval with Irigaray," *Hypatia,* 15(3): 73–91.

Olsen, G. *et al.* (1998) "An Epidemiologic Investigation of Reproductive Hormones in Men with Occupational Exposure to Perfluorooctanoic Acid," *JOEM,* 40(7): 614–22.

Orbach, S. (1986) *Hunger Strike: The Anorexic's Struggle as a Metaphor for Our Age.* London: Faber.

Oring, L., Fleischer, R., Reed, J. and Marsden, K. (1992) "Cuckoldry Through Stored Sperm in the Sequentially Polyandrous Spotted Sandpiper," *Nature,* 359: 631–3.

Oudshoorn, N. (1990) "On the Making of Sex Hormones: Research Materials and the Production of Knowledge," *Social Science Studies,* 20: 5–33.

Oudshoorn, N. (1994) *Beyond the Natural Body. And Archaeology of Sex Hormones.* London and New York: Routledge.

Overall, C. (1989) *The Future of Human Reproduction.* Toronto: The Women's Press.

Owen, D.F. (1988) "Mimicry and Transvestism in Papilio Phorcas," *Journal of Entomological Society of Southern Africa*, 51: 294–6.

Oyama, S. (2000) *The Ontogeny of Information: Developmental Systems and Evolution.* Durham, NC: Duke University Press.

Pagon, R.A. (1987) "Diagnostic Approach To The Newborn With Ambiguous Genitalia," *Pediatric Clinics of North America* 34: 1019–31.

Palca, J. (1990) "The Other Human Genome" Science, 249: 1104–1105.

Parker, G. and Pearson, R. (1976) "A Possible Origin and Adaptive Significance of the Mounting Behavior Shown by Some Female Mammals in Oestrus," *Journal of Natural History*, 10: 241–5.

Patten, B. (1997) "On Science, Identity Politics, and Group-Speak" in P. Gowaty (ed.) *Feminism and Evolutionary Biology. Boundaries, Intersections, and Frontiers.* New York: Chapman and Hall, pp. 562–8.

Paulozzi, L. *et al.* (1997) "Hypospadias Trends in Two American Surveillance Systems," *Pediatrics*, 100: 831–4.

Pavelka, M.M. (1995) "Sexual Nature: What Can We Learn from a Cross-Species Perspective?" in P. Abramson and S. Pinkerton (eds) *Sexual Nature, Sexual Culture.* Chicago, IL: University of Chicago Press, pp. 17–36.

Pearson, H. (2002) "Dual Identities," *Nature*, 417, 2 May, pp. 10–11.

Perlmutter, A.D. and Reitelman, M.D. (1992) "Surgical Management of Intersexuality" in P.C. Walsh, A.B. Retik, T.A. Stamey, and E.D. Vaughan (eds) *Campbell's Urology.* Philadelphia, PA: Saunders, pp. 1951–66.

Peterson, A. (1998) "Sexing the Body: Representations of Sex Differences in Gray's *Anatomy*, 1858 to the Present," *Body and Society*, 4(1): 1–15.

Plant, S. (1997) *Zeros and Ones.* London: Fourth Estate.

Plotnitsky, A. (1994) *Complementarity: Anti-Epistemology after Bohr and Derrida.* Durham, NC: Duke University Press.

Policansky, D. (1982) "Sex Change in Plants and Animals," *Annual Review of Ecology and Systematics*, 13: 471–95.

Poole, F.J.P. (1996) "The Procreative and Ritual Constitution of Female, Male and Other: Androgynous Beings in the Cultural Imagination of the Bimin-Kuskusmin of Papua New Gunea," in S.P. Ramet (ed.) *Gender Reversals and Gender Cultures: Anthropological and Historical Perspectives.* London: Routledge, pp. 197–218.

Preves, S. (2003) *Intersex and Identity. The Contested Self.* New Brunswick, NJ and London: Rutgers University Press.

Prigogine, I. and Stengers, I. (1984) *Order Out of Chaos: Man's New Dialogue With Nature.* New York: Bantam.

Rabinow, P. (1992) "Artificiality and Enlightenment: From Sociobiology to Biosociality," in J. Crary and S. Kwinter (eds) *Incorporations.* New York: Urzone Books, pp. 234–53.

Rabinow, P. (1997) *Michel Foucault: Ethics, Subjectivity and Truth. The Essential Works of Foucault 1954–1984.* Vol. 1. New York: The New Press.

Rackham, M. (2000) "My Viral Lover," *RealTime/OnScreen*, 37 June: 22.

Raloff, J. (1994) "That Feminine Touch: Are Men Suffering from Prenatal or Childhood Exposures to 'Hormonal Toxicants'?" *Science News*, 145(4): 56–8.

Reeder, A. *et al.* (1998) "Forms and Prevalence of Intersexuality and Effects of Environmental Contaminants on Sexuality in Cricket Frogs," *Environmental Health Perspectives*, 106(5): 261–6.

Reinboth, R. (ed.) (1975) *Intersexuality in the Animal Kingdom*. New York and Heidelberg Berlin: Springer-Verlag.

Reite, M. and Caine, N. (eds) (1983) *Child Abuse: The Nonhuman Primate Data*. New York: Alan R. Liss Inc.

Ridley, M. (2003) *Nature Via Nurture. Genes, Experience and What Makes Us Human*. London: Fourth Estate.

Roberts, C. (1999) "Thinking Biological Materialities," *Australian Feminist Studies*, 14(29): 131–9.

Roberts, C. (2003a) "Drowning in a Sea of Estrogens: Sex, Hormones, Sexual Reproduction and Sex" *Sexualities*, 6(2): 195–213.

Roberts, C. (2003b) "Emasculating Hormones in the 21st Century." *Paper presented at the Vital Politics*, BIOS, London, 7–9 September 2003.

Rose, N. (1996) *Inventing Ourselves*. Cambridge, MA: Cambridge University Press.

Rothblatt, M. (1995) *The Apartheid of Sex*. New York: Crown Publishers.

Rowell, T. (1974) "The Concept of Social Dominance," *Behavioral Biology*, 11: 131–154.

Rowell, T. (1979) "How Would We Know if Social Organization Were Not Adaptive?," in I. Bernstein and E. Smith (eds) *Primate Ecology and Human Origins*. New York: Garland, pp. 1–22.

Rowell, T. (1984) "Introduction: Mothers, Infants and Adolescents," in M. Small (ed.) *Female Primates*. New York: Alan Liss, pp. 13–16.

Ruse, M. (1981) "Are There Gay Genes? Sociobiology and Homosexuality," *Journal of Homosexuality*, 6: 5–34.

Ruse, M. (1988) *Homosexuality: A Philosophical Inquiry*. Oxford: Blackwell.

Sagan, D. (1992) "Metametazoa: Biology and Multiplicity," in J. Crary and S. Kwinter (eds) *Incorporations*. New York: Urzone Books, pp. 362–85.

Sangha, K., Stephenson, M. Brown, C. and Robinson, W. (1999) "Extremely Skewed X-Chromosome Inactivation is Increased in Women with Recurrent Spontaneous Abortion," *American Journal of Human Genetics*, 65: 917–21.

Sayers, J. (1982) *Biological Politics: Feminist and Anti-Feminist Perspectives*. London: Tavistock Publishing.

Santti, R. *et al.* (1998) "Phytoestrogens: Potential Endocrine Disruptors in Males," *Toxicology and Industrial Health*, 14(1–2): 223–37.

Sawicki, J. (1999) "Disciplining Mothers: Feminism and the New Reproductive Technologies," in J. Price and M. Shildrick (eds) *Feminist Theory and the Body*. Edinburgh: Edinburgh University Press.

Schatten, G. and Schatten, H. (1983) "The Energetic Egg," *The Sciences*, September/October: 28–34.

Schiebinger, L. (1989) *The Mind Has No Sex? Women in the Origins of Modern Science*. Cambridge, MA: Harvard University Press.

Schiebinger, L. (1993) *Nature's Body*. London: Pandora.

Schiebinger, L. (1999) *Has Feminism Changed Science?* Cambridge, MA: Harvard University Press.

Sedgwick, E. (1991) *Epistemology of the Closet*. New York: Harvester Wheatsheaf.

Sedgwick, E. (1995) *Shame and its Sisters*. Durham, NC: Duke University Press.

Schneider, D. (1968) *American Kinship: A Cultural Account.* Englewood Cliffs, NJ: Prentice Hall.

Schneider, D. (1980) *American Kinship: A Cultural Account.* 2nd Edition. Chicago, IL: University of Chicago Press.

Seidman, S. (1994) *Contested Knowledge: Social Theory in the Postmodern Era.* Oxford: Blackwell Publishers.

Shaw, E. and Darling, J. (1985) *Female Strategies. Two Scientists Investigate the Varieties of Sexual Behavior in the Animal World and Revolutionize the Concept of Femininity.* New York: Simon and Schuster, Inc.

Sheridan, S. (2002) "Words and Things: Some Feminist Debates on Culture and Materialism," *Australian Feminist Studies*, 17(37): 23–30.

Shildrick, M. (1996) "Posthumanism and the Monstrous Body," *Body and Society*, 2(1): 1–15.

Skakkeæk, N. *et al.* (1998) "Germ Cell Cancer and Disorders of Spermatogenesis: An Environmental Connection?" *APMIS*, 106: 3–12.

Small, M. (ed.) (1984) *Female Primate: Studies by Female Primatologists.* New York: Liss.

Small, M. (1993) *Female Choices: Sexual Behavior of Female Primates.* Ithaca, NY: Cornell University Press.

Smith, C.L. (1967) "Contribution to a Theory of Hermaphroditism," *Journal of Theoretical Biology*, 17: 76–90.

Smith, M. and Szathmáry, E. (1995) *The Major Transitions in Evolution.* Oxford: W.H. Freeman Spektrum.

Snowdon, C. (1997) "The 'Nature' of Sex Differences: Myths of Male and Female," in P. Gowaty (ed.) *Feminism and Evolutionary Biology. Boundaries, Intersections, and Frontiers.* New York: Chapman and Hall, pp. 276–93.

Soper, K. (1995) *What is Nature?* Oxford and Cambridge, MA: Blackwell.

Sork, V. (1997) "Quantitative Genetics, Feminism, and Evolutionary Theories of Gender Differences," in P. Gowaty (ed.) *Feminism and Evolutionary Biolology. Boundaries, Intersections, and Frontiers.* New York: Chapman and Hall, pp. 86–115.

Spallone, P. and Steinberg, D. (1987) *Made to Order: The Myth of Reproductive and Genetic Progress.* New York: Pergamon Press.

Spanier, B. 1991. " 'Lessons' from 'Nature': Gender Ideology and Sexual Ambiguity in Biology," in K. Straub and J. Epstein (eds) *Body Guards: The Cultural Politics of Gender Ambiguity.* New York: Routledge, pp. 329–50.

Spanier, B. (1995) *Im/Partial Science.* Bloomington, IN: Indiana University Press.

Sprin, A.W. (1984) *The Granite Garden: Urban Nature and Human Design.* New York: Basic Books.

Stengers, I. (1997) *Power and Invention.* Minneapolis, MN: University of Minnesota Press.

Stengers, I. (2000) "Another Look: Relearning to Laugh," *Hypatia*, 15(4): 41–54.

Stone, S. (1991) "The Empire Strikes Back: A Post-Transsexual Manifesto," in J. Epstein and K. Straub (eds) *Body Guards: The Cultural Politics of Gender Ambiguity.* New York: Routledge, pp. 280–304.

Strain, L., Warner, J., Johnston, T., and Bonthron, D. (1995) "A Human Parthenogenetic Chimaera," *Nature Genetics*, 11 October: 164–9.

Strain, L., Dean, J., Hamilton, M., and Bonthron, D. (1998) "A True Hermaphrodite Chimera Resulting from Embryo Amalgamation after In Vitro Fertilization," *The New England Journal of Medicine*, 338(3): 166–9.

Strathern, M. (1980) "No Nature, No Culture: The Hagen Case," in Carol MacCormack and Marilyn Stratern (eds) *Nature, Culture and Gender*. Cambridge, MA: Cambridge University Press, pp. 174–222.

Strathern, M. (1992) *Reproducing the Future: Anthropology, Kinship and the New Reproductive Technologies*. Manchester: Manchester University Press.

Subramanian, S. (1995) "The Story of Our Genes," *Time*, 16 January, pp. 54–5.

Szasz, T. (1970) *The Manufacture of Madness*. New York: Harper and Row.

Taylor, E. (1871/1927) *Primitive Culture*. London: John Murray.

The Biology and Gender Study Group (1989) "The Importance of Feminist Critique for Contemporary Cell Biology," in N. Tuana (ed.) *Feminism and Science*. Bloomington, IN: Indiana University Press, pp. 172–87.

Thomas, L. (1974) *The Lives of a Cell*. New York: The Viking Press.

Thompson, C. (2001) "Strategic Naturalizing: Kinship in an Infertility Clinic," in S. Franklin and S. McKinnon (eds) *Relative Values. Reconfiguring Kinship Studies*. Durham, NC: Duke University Press, pp. 175–202.

Thornhill, R. (1980) "Rape in *Panorpa* scorpionflies and a General Rape Hypothesis," *Animal Behavior*, 28: 52–9.

Tobach, E. and Rosoff, B. (eds) (1978) *Genes and Gender*. New York: Gordian Press.

Triea, K. 1996. "Untitled" : http://www.qis.net/~triea/kira.html.

Tuana, N. (1989) "The Weaker Seed: The Sexist Bias of Reproductive Theory," in N. Tuana (ed.) *Feminism and Science*. Bloomington, IN: Indiana University Press, pp. 147–71.

Turner, B. (1987) *Medical Power and Social Knowledge*. London: Sage.

Twain, M. (1961) *Life on the Mississippi*. New York: Signet Books.

Tyler, C. *et al.* (1998) "Endocrine Disruption in Wildlife: A Critical Review of the Evidence," *Critical Reviews of Toxicology*, 28(4): 319–61.

Tyler, P. (1984) "Homosexual Behavior in Animals," in K. Howells (ed.) *The Psychology of Sexual Diversity*. Oxford: Blackwell, pp. 42–62.

Van Dijk, B., Boomsma, D., and de Man, A. (1996) "Blood Group Chimerism in Human Multiple Births is Not Rare," *American Journal of Medical Genetics*, 61: 264–8.

Van Loon, J. (2000) "Parasite Politics: On the Significance of Symbiosis and Assemblage in Theorizing Community Formations," in C. Pierson and S. Tormey (eds) *Politics at the Edge*. New York: St. Martin's Press.

Vasey, P. (1995) "Homosexual Behavior in Primates," *International Journal of Primatology*, 16: 173–204.

Waage, J. (1997) "Parental Investment – Minding the Kids or Keeping Control?" in P. Gowaty (ed.) *Feminism and Evolutionary Biology. Boundaries, Intersections, and Frontiers*. New York: Chapman and Hall, pp. 527–53.

Waage, J. and Gowaty, P. (1997) "Myths of Genetic Determinism," in P. Gowaty (ed.) *Feminism and Evolutionary Biology. Boundaries, Intersections, and Frontiers*. New York: Chapman and Hall, pp. 585–614.

Waldby, C. (1999) "IatroGenesis: The Visible Human Project and the Reproduction of Life," *Australian Feminist Studies*, 14(29): 77–90.

Wallen, K. (1995) "The Evolution of Female Sexual Desire," in P. Abramson and S. Pinkerton (eds) (1995) *Sexual Nature, Sexual Culture*. Chicago, IL: University of Chicago Press, pp. 57–79.

Ward, J. (1987) "The Nature of Heterosexuality," in G.E. Hanscombe and M. Humphries (eds) *Heterosexuality*. London: GMP Publishers, pp. 145–69.

Warner, R.R. (1984) "Mating Behavior and Hermaphroditism in Coral Reef Fish," *American Scientist*, 72: 128–36.

Warner, R.R. (1975) "The Adaptive Significance of Sequential Hermaphroditism in Animals," *American Naturalis*, 109: 61–82.

Wasser, S.K. and Barash, D.P. (1983) "Reproductive Suppression Among Female Mammals: Implications for Biomedicine and Sexual Selection Theory," *Quarterly Review of Biology*, 58: 513–38.

Weinrich, J.D. (1976) *Human Reproductive Strategy: The Importance of Income Predictability, and the Evolution of Non-reproduction.* Unpublished Doctoral Dissertation, Harvard University.

Weinrich, J.D. (1980) "Homosexual Behavior in Animals: A New Review of Observations form the Wild, and Their Relationship to Human Sexuality," in R. Forleo and W. Pasini (eds) *Medical Sexology: The Third International Congress.* Littleton, MA: PSG Publishing, pp. 288–95.

Weinrich, J.D. (1982) "Is Homosexuality Biologically Natural?" in W. Paul, J.D. Weinrich, J.C. Gonsiorek, and M.E. Hotveldt (eds) *Homosexuality: Social, Psychological, and Biological Issues.* Beverly Hills, CA: Sage Publications, pp. 197–211.

Weinstein, J. (2003) "Traces of the Beast: Becoming-Nietzsche, Becoming-Animal, and the Figure of the Trans-Human," in Rowman and Littlefield (eds) *Nietzsche's Bestiary* (forthcoming).

Weinstein, J. (2004) "Traces of the Beast: Becoming Nietzsche, Becoming Animal, and the Figure of the Transhuman" in C.D. Acampora and R.R. Acampora (eds) *A Nietzchean Bestiary: Becoming Animal Beyond Docile and Brutal.* Lanham, Maryland: Rowman and Littlefield Publishers, Inc., pp. 301–18.

Weir, L. (1998) "Pregnancy Ultrasound in Maternal Discourse," in M. Shildrick and J. Price (eds) *Vital Signs.* Edinburgh: Edinburgh University Press, pp. 78–101.

Weston, K. (2001) "Kinship, Controversy, and the Sharing of Substance: The Race/Class Politics of Blood Transfusion," in S. Franklin and S. McKinnon (eds) *Relative Values. Reconfiguring Kinship Studies.* Durham, NC: Duke University Press, pp. 147–74.

Wertheim, M. (1995) *Pythagoras' Trousers: God, Physics and the Gender Wars.* New York: Random House.

Whittle, S. (1998) "The Trans-cyberian Mail Way," *Social and Legal Studies*, 7: 389–408.

Whittle, S. (2000) "The Becoming Man: The Law's Ass Brays," in K. More and S. Whittle (eds) *Reclaiming Genders.* London: Cassette College Audio, pp. 15–33.

Wickler, W. (1967) "Socio-sexual Signals and Their Intra-specific Imitation Among Primates," in D. Morris (ed.) *Primate Ethology.* London: Weidenfield and Nicolson, pp. 69–147.

Wieser, W. (1997) "A Major Transition in Darwinism," *Trends in Ecology and Evolution*, 12: 367–70.

Williams, G. (1975) *Sex and Evolution.* Princeton, NJ: Princeton University Press.

Willmott, J. (2000) "The Experiences of Women with Polycystic Ovarian Syndrome," *Feminism and Psychology*, 10: 107–16.

Willson, M.F., and E.R. Pianka (1963) "Sexual Selection, Sex Ratio, and Mating Systems," *American Naturalist*, 97: 405–7.

Wilson, E. (1996) "On the Nature of Neurology," *Hysteric Body and Medicine*, 49–63.

Wilson, E. (1998) *Neural Geographies. Feminism and the Microstructure of Cognition.* New York and London: Routledge.

Wilson, E. (1999) "Melancholic Biology: Prozac, Freud, and Neuronal Determinism," *Configurations*, 7(3): 403–19.

Wilson, E. (2000) "Scientific Interest: Introduction to Isabelle Stengers, 'Another Look: Relearning to Laugh,' " *Hypatia*, 15(4): 38–40.

Wilson, E.O. (1975) *Sociobiology: The New Synthesis.* Cambridge, MA: Harvard University Press.

Wilson, E.O. (1978) *On Human Nature.* Cambridge, MA.: Harvard University Press.

Wilson, E.O. (2000) *Sociobiology: The New Synthesis. 25th Anniversary Edition.* Cambride, MA.: Harvard University Press.

Wittig, M. (1993) "One is Not Born a Woman," in H. Abelove, M. Barale, and D. Halperin (eds) *The Lesbian and Gay Studies Reader.* New York: Routledge, pp. 103–9.

Witz, A. (2000) "Whose Body Matters? Feminist Sociology and the Corporeal Turn in Sociology and Feminism," *Body and Society*, 6(2): 1–24.

Yokoya, S., Kato, K., and Suwa, S. (1983) "Penile and Clitoral Size in Premature and Normal Newborn Infants and Children," *Horumon To Rinsho*, 31: 1215–20.

Zakin, E. (2000) "Bridging the Social and the Symbolic: Toward a Feminist Politics of Sexual Difference," *Hypatia*, 15(3): 19–44.

Zihlman, A. (1985) "Gathering Stories for Hunting Human Nature," *Feminist Studies*, 11: 364–77.

Zola, I.K. (1972) "Medicine as an Institution of Social Control," *Sociological Review*, 20: 487–504.

Index